SUPERDOVE

DOVE

How the Pigeon

Took Manhattan . . .

and the World

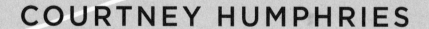

COURTNEY HUMPHRIES

Smithsonian Books

COLLINS
An Imprint of HarperCollins Publishers

HarperCollins books may be purchased for educational, business, or sales promotional use. For information, please write: Special Markets Department, HarperCollins Publishers, 10 East 53rd Street, New York, NY 10022.

FIRST EDITION

Book design by Sunil Manchikanti

Printed on acid-free paper

Library of Congress Cataloging-in-Publication Data

Humphries, Courtney, 1977-
    Superdove : how the pigeon took Manhattan . . . and the world / Courtney Humphries. — 1st ed.
        p. cm.
    ISBN 978-0-06-125916-6
    1. Pigeons—Anecdotes. I. Title.
    QL696.C63H86   2008
    598.6'5—dc22

                                2008033386

08 09 10 11 12 OV/RRD 10 9 8 7 6 5 4 3 2 1

FOR JAMES

Contents

Acknowledgments

This book grew out of an assignment in MIT's Graduate Program in Science Writing, and I'm grateful for the support and feedback of the program's faculty as I developed my ideas into a book, particularly Marcia Bartusiak and Robert Kanigel. My agent, Russell Galen, supported this book from its nascence and gave me guidance in shaping my ideas into a longer narrative. I'm also grateful for the feedback I received from my editors at Smithsonian Books, T. J. Kelleher and Elisabeth Dyssegaard.

Many chapters involved digging for pigeon information in libraries, and I had help in my efforts from the staff at Harvard's libraries: the Widener Library, the Ernst Mayr Library of the Museum of Comparative Zoology, and the Harvard Archives. As I researched pigeons, I was fortunate that many scientists and pigeon-lovers shared their time and thoughts with me. My early discussions with Richard Johnston and Louis Lefebvre were instructive in shaping the focus of this book. I'm grateful to other scientists who spent time talking with me about pigeons, bird flight and navigation, and urban ecology, and especially those whose research did not directly involve pigeons but were willing to help me draw connections and place pigeons in a larger scientific context. Johnston, Robert Cook, and Daniel Haag-Wackernagel were kind to spend extended time meeting with me. Emilio Baldaccini and Marco Apollonio helped my visit to Sardinia go smoothly, and I'm especially indebted to graduate students Ila Geigenfeind, Antonio Cossu, and Laura Iacolina for the time and cheerful effort they spent showing me their local pigeon populations.

I benefited from the insights of several people into the management and ecology of urban pigeons and their place in cities. The pigeon activists of New York were very forthcoming and helpful, particularly Al Streit and Johanna Clearfield. And I would have had a difficult time navigating the world of pigeon breeding without the fanciers who were very generous with their time and information, including Joe and Michelle Cussick and others not mentioned in the text.

I faced many reactions when I told people I was writing a book about pigeons, from fascination to complete skepticism. But ultimately I received a great deal of support from friends and family members who were always sending me pigeon-related news clippings and links, which I'm very thankful for. I even learned to appreciate people telling me: "Oh, I saw some pigeons the other day and thought of you." I'm grateful for the support of my parents, Sandra and Stanley, and my brother, Colin. And most of all for the support of my husband, James, for patiently reading and editing my work, joining me for urban walks and pigeon shows, and never balking when I suggested pigeon-watching on our honeymoon.

SUPERDOVE

1

The Pigeon's Progress

I don't know what it is about fecundity that so appalls. I suppose it is the teeming evidence that birth and growth, which
we value, are ubiquitous and blind, that life itself is so astonishingly cheap, that nature is as careless as it is bountiful, and
that with extravagance goes a crushing waste that will one day
include our own cheap lives.

ANNIE DILLARD

I first really noticed pigeons when I traveled in Europe after college.
I sat in the Piazza San Marco in Venice, where it was impossible
not to notice them. They gathered in a gray, shuddering mass, an
ocean of bodies at once marvelous and disgusting. Children, I noticed,
tended to see what was marvelous, to step into that vibrating pool of
pigeons and watch the chaos as the birds parted around them. Adults
were more on the side of disgust; they edged away from the birds and
thought of crowds and filth and disease. But in spite of its sanitary implications, the sight of that throng of pigeons was certainly arresting, a
muddled swarm of life in an otherwise genteel old city.

I had many opportunities to watch pigeons as a tourist that summer.
I noticed how the pigeons moved, how the males puffed themselves
up and drove the females in circles around the piazzas of Italy, and in
London's squares, and in the public gardens of Paris. Amid the bustle
of these great cities, pigeons were carrying on their lives, feeding and
procreating as resolutely as any of the human residents.

A few years later, I took a trip with two friends to Thailand; it was my first visit to Asia and the farthest I had ever traveled. Walking across a plaza in Bangkok on the first day, we ran into a celebration for the Queen's birthday. Her portrait soared over a soundstage, where children in bright outfits performed traditional Thai dances for a small crowd. It was the sort of scene a traveler loves to stumble upon. But in the background I noticed something else: groups of slate-gray pigeons wandering through the square. There is a particular kind of disappointment a traveler feels upon realizing how similar one city is to another. Not only is crossing the globe a simple matter of hours spent on a plane, but the cities are adorned with some of the same banks and fast-food restaurants, the same universal traffic signs, the same modern buildings and movie theaters. And, in this case, the same birds. I began to think of pigeons not just as companions on my travels but as symbols of homogeneity.

Later, my conception of pigeons as identical gray blobs populating the planet was challenged by reading Charles Darwin's *Origin of Species*. I was surprised to find that he chose to begin that famous book with a lengthy chapter about pigeons; for some reason, Darwin had used these birds to illustrate an argument about the incredible diversity of life, ascribing to pigeons a complexity I had never considered before.

I had never learned a thing about pigeons in school or watched a documentary about them or read about them in books. Pigeons were a fact I had taken for granted. They were background scenery or extras in movies: so common they were invisible. They were at once familiar and completely unknown. Inspired by Darwin, I began to investigate where these birds came from and how they came to appear in such abundance in every city I ever visited; this book is the result of rethinking pigeons.

Pigeons have long been familiar to people, but their image has not always been as a ubiquitous urban pest. Their Latin name, *Columba livia*, means a dove the color of lead. At one time, the words "pigeon" and "dove" were used interchangeably; once you know that, you realize how very different *Columba livia*'s image once was. There is good evidence that much of the iconic imagery of doves we still recognize today was originally based on the same species as our pesky street pigeons, *C. livia*. The British historian Jean Hansell has published books exhaustively listing the various appearances of pigeons in iconography of the past. They were fertility symbols associated with the ancient goddesses Ishtar, Aphrodite, and Venus. They are the most frequently mentioned birds in the Old Testament, where they were often used as sacrifices and messengers. In Christianity, doves came to symbolize the Holy Spirit itself.

But "pigeon" and "dove" gradually came to acquire very different meanings. Perhaps Shakespeare is to blame; he almost always used "pigeon" and "dove" in different ways. Pigeons appeared in practical roles, as food, letter-carriers, and sometimes as symbols of fidelity and care of children. They were also known for their unusual digestive systems that lacked a gall bladder; thus Hamlet castigates himself for being "pigeon-liver'd" for his want of gall. Doves, however, were equated with peace, modesty, patience, love, and other noble ideals. Since then, the meanings of "pigeon" and "dove" have only grown farther apart. A dove, not a pigeon, brought Noah an olive leaf. We never talk of pigeons of peace or dove droppings on statues. "Dove" is a pleasant enough title to grace chocolate bars and soap, while "pigeon" has no marketing appeal.

While the images of doves continue to decorate Christmas cards, the real birds that once inspired that image have been cast out of such heavenly associations. But even as city birds, pigeons were once romantic; old photographs of urban scenes often show pigeons flying through streets, denizens of a new world of bridges and train stations and smokestacks. But just as big cities have lost some of their romance, so have pigeons. When Woody Allen famously called them "rats with wings" in *Stardust Memories*, their lot as urban pests was solidified. At best, pigeons appear these days in *New Yorker* cartoons as cheeky city dwellers; they brave traffic and ride the subways and befriend old ladies on benches. At worst, they are filthy poop machines that spread disease, and freeloaders that scrounge through trash for their meals. They drain money from cities; industries have emerged to help people repel and kill them. No longer a lofty abstraction, pigeons have become all too real, a constant presence of flesh and feathers and shit. Doves are something pure and lovely, pigeons the rough reality.

Nevertheless, until recently the pigeon had one last claim to dovish gentility; in English, the common name of *Columba livia* has for centuries been Rock Dove. But in 2003, the American Ornithologist's Union announced, in its annual *Checklist of Birds*, that the official name would henceforth be Rock Pigeon. The chairman of the AOU's committee on classification and nomenclature told me that rock doves were one of just two species in their genus that were called doves, whereas most other doves are in the genus *Streptopelia*. When people think of doves, they generally think of a bird that is slender and long like the Mourning Dove, not short and squat like *Columba livia*. "It's a pigeon in every sense of the word," he said. And so, long after it had ceased to seem dovelike, *Columba livia* is just a pigeon.

This symbolic and linguistic history of pigeons has all the outlines of a fall from grace. It's hard to imagine how an animal's status in our society could topple farther. But the physical reality of pigeons tells a different story. What pigeons mean to us, after all, is very different from what it means to be a pigeon. The pigeons that I saw on my travels are one and the same species. They probably originated somewhere in the Middle East hundreds of thousands of years ago, but today they blanket the earth, concentrated largely in cities and towns. They are among the most recognized and abundant birds. Clearly, from a pigeon's perspective, their story is not a tragedy.

Though *C. livia* is the most common of its kind, others species of pigeons and doves are also on the move, taking up new territories and becoming more abundant. In his book *Pigeons and Doves of the World*, Derek Goodwin notes that pigeons as a group are "highly edible" birds that have few defenses, but are surprisingly successful in spite of it. He attributes this success to "being physically and psychologically 'tough' in spite of their delicate, fragile appearance and their timidity." They may be easy prey, but pigeons are able to out-reproduce their hunters and survive in many different environments. Their gentleness should not be mistaken for delicacy. In their quiet way, pigeons and doves have managed to succeed very well in vastly different conditions and habitats.

The ultimate demonstration of their success is the swarms of street or "feral" pigeons that fill our cities. Early in my search for information on pigeons, I came across a scientific book devoted solely to these ubiquitous birds. Its primary author, an ornithologist from Kansas named Richard Johnston, made a comment about how we think about pigeons, which stood out amidst the more dry scientific language. "The special qualities of feral pigeons are rarely recognized as special, which is a result of the way humans perceive the natural world," he wrote. "Dominant western worldviews have taught that nature exists for human use and that humans are its custodians or curators, fundamentally apart from the natural world. This philosophic position has been unprofitable in many ways, one of which is important here: Because humans think of their activities as different from 'nature,' they are deemed artifacts, derived from human skills—not natural."

Here was a bold statement. Perhaps pigeons were not inherently boring as I had assumed; instead, perhaps our blind spots keep us from appreciating them. Our disgust blinds us to any living thing so abundant as those birds in Venice, that fecundity we find so appalling. When something is everywhere, it paradoxically becomes invisible and its value diminishes in our minds. Johnston's words suggest that our prejudices prevent us from looking at an urban animal like the pigeon as a marvel of nature, and not simply a pest. He seemed to believe that pigeons are not only interesting, but that understanding them better could, in its own way, offer an indictment of our entire worldview about nature. By ignoring pigeons, perhaps we were missing an opportunity to hear a different kind of story, a story of success in a changing world.

2

Invited Guests

It's a Sunday afternoon in winter, and Boston has warmed to above-freezing temperatures after a long cold spell. From the view in Boston Common, the city's large downtown park, you'd think spring had arrived. Shoppers from nearby department stores mill around the bronze fountain. Dog walkers cross the park on paved paths edged in snow. Men and groups of teenagers sit on benches in the sun. A patch of matted grass nearby is awash with pigeons, eager to take advantage of this sudden show of people, who always mean food. Red feet splayed on the hard ground, they peck resolutely, looking for crumbs or seeds or whatever is edible.

The scene is not unique to Boston. Similar arrangements of people and pigeons fill urban parks across the country at any time of the year. Often the pigeons will be joined by bands of sparrows and starlings, searching for food together across the blocks of lawn and stripes of pavement. They rule the parks, this trinity of urban bird life.

Being abundant doesn't win you much respect, and pigeons, starlings, and sparrows are particularly maligned. They each get dutiful but apologetic entries in bird guides, as if to say, "these are the birds you always see but don't really care about—neither do we." They are among a handful of bird species not protected by federal law, and they are often called nuisances, pests, and worst of all, invasive species. Yes, this trio of ubiquitous birds, the ones so common we barely notice them, never existed in North America before Europeans came. They don't belong.

Invasive species have become the pariahs of ecology—they change ecosystems, choke other species out of existence, and generally threaten regional uniqueness. Educational campaigns have been launched to help people understand that there is a new hierarchy of life: native and invasive. The title invasive species suggests that some inherent quality impels it to maraud like an army or slink like a thief. It blames the species itself. But the vast majority of invasions happen because humans make them possible.

The perfect invasive species—one that is mobile, spreads quickly, modifies its ecosystem, and drives other species to extinction—is our own. Humans are the most powerful invaders on the planet, and many of the other species that we call aliens or invasives are merely adept at taking advantage of the opportunities we create. A plant might not act as an invader in its own territory where it is held within ecological checks and balances, but when moved by humans it flourishes beyond control. Other introduced species, such as the quail—a half million of which were brought into the U.S. between 1875 and the 1950s—fail to establish self-sustaining populations.

Each invasive species has a story behind it, a set of conditions that brought it to its adopted home. Starlings were part of an ill-conceived effort to introduce all the birds of Shakespeare into Central Park in the 1850s. At the time, it must have seemed harmless to release one hundred of these fetching black birds into a green area safely enclosed by Manhattan. But 150 years later they had reached the Pacific and Alaska and their total population had soared to 200 million. House sparrows were also brought into nearby Brooklyn in 1851, and again later in western states. They too spread across the continent. These cases can be seen as the isolated acts of a few shortsighted individuals. But the introduction of pigeons to North America was a collective process that took place over centuries.

Several yards away from where I sit in the Common, two men arrive and begin to toss hunks of bread to the pigeons. At first a few birds gather around them, then more stream in as they clue in to the feast. Although they are intent on eating, they are also on alert. Every minute or so, some sound or movement launches the whole flock into the air in a shudder. Each time they take off, the pigeons swoop low over the benches near the fountain, nearly brushing the heads of those of us sitting. A couple of teenage girls next to me scream and hug each other at each fly-by. I try to hold my ground, sitting upright as the birds bullet toward me, wings flapping—but I lose, ducking as they pass.

It's impossible to understand how pigeons took over this land without delving into the story of their relationship with people. Today we know pigeons as vermin, as benign neighbors in a city, or as an attraction to feed and watch. But whether the other people in Boston Common see pigeons as wildlife or pests, cute or hideous, whether they prefer to give them bread or a furtive kick once in a while, no one looks at these pigeons and thinks "dinner." Even the men feeding the pigeons don't expect a meal in return.

And yet, for thousands of years, pigeons have been kept by people for food. Unlike sparrows and starlings, pigeons were never simply decorative. They were brought here to feed us, not to populate parks. So if pigeons crowd our cities now, we have the appetites of our ancestors to blame.

Thousands of years ago, rock pigeons lived wild on sea cliffs in the Middle East and parts of Europe. They tucked nests into the crevices of rocks, gathered on the cliffs in colonies, dodged the sharp eyes of falcons. For the most part, these birds stayed put. They didn't migrate south for the winter nor move from place to place. Instead, they made trips over land by day to find seeds and berries, returning to their nesting grounds each night. Because they had relatively homebound lives and they chose to roost only in the most inaccessible rocks and cliffs, anyone who considered the birds would probably not have thought them candidates for worldwide conquest.

But then populations of humans in the Middle East began to adopt a new lifestyle that changed everything. Once people created permanent settlements, they began to transform the landscape. Rock dwellers, pigeons were confined to relatively dry, treeless areas. But people built structures of mud and stone that they called houses and temples but made lovely cliffs. Pigeons were able to move into new territories as humans created these new habitats for them. And unlike the cliffs they came from, these had few resident falcons. It's not hard to imagine, looking at pigeons alight on buildings in cities, that they may have chosen us just as much as we chose them. The traditional story of domestication is "man conquered beast." Humans, the story goes, went out into the wild, brought baby animals back with them, and raised them in captivity until they were eventually "tamed." Certainly people in early civilizations may have taken young pigeons from their nest and kept them in cages until the birds were used to being cared for by humans. But the reality is more complex.

Pigeons are likely the first bird to be domesticated—sometime around 3000 BC. They were kept for food and also as pets; they were sometimes associated with goddesses and fertility and viewed as sacred animals. They supplied us with meat and served as cultural symbols. But pigeons also had something to gain from domestication—food, shelter, and protection from predators. Domestication is less a one-sided relationship and more a process of coevolution.

However they came to be a part of society, pigeons soon found a place at the table. Biblical stories refer to them as sacrificial offerings, suggesting they were commonly eaten. The first evidence of large-scale domestication comes from Egypt, where Pharaoh Ramses II offered over 57,000 pigeons to the god Amon. To capture and collect wild rock pigeons living hundreds of miles from the offering site would have been an enormous task, so the Egyptians were probably breeding their own stocks at this time. And large-scale breeding would have required official housing for the birds.

Given the great effort and expense that people expend trying to keep pigeons off buildings today, it's ironic that for most of history we have been building structures to lure pigeons to stay. An entire architecture of dovecotes and lofts developed around the breeding of pigeons. Some of the earliest lures were simple clay pots stuck together or hung outside a building. But soon specific buildings were dedicated to pigeons. In Egypt, dovecotes were made of mud and had cone-shaped roofs with holes through which the pigeons could fly in and out. Excavations of an Egyptian farming town in the Roman Empire, Karanis, revealed several pigeon houses, suggesting a large-scale operation—one building contained three large towers where pigeons lived. Grain was kept at the bottom, and the walls were lined with clay pots for nests.

With the Romans, pigeon-keeping spread in Europe, where housing for pigeons was added on to the upper story of buildings and in towers and turrets. In Renaissance Italy, houses were often built with a small belvedere or open tower at the top capped with a hipped roof where pigeons lived, a style that eventually became a standard feature of the Italian country villa. Pigeons were also kept in free-standing buildings that housed hundreds of birds. Dovecotes became a common feature of the countryside in Italy, France, and England. Even at the end of the 1700s, when the use of pigeons for food was declining in England, dovecotes remained popular as the showpiece of every fashionable estate. Architects drew on contemporary romantic tastes; some dovecotes looked like classical temples, others had cupolas or spires, or were paired together and joined by an arch. For the country retreat of Wroxton Abbey in Oxfordshire, the owners commissioned a Gothic dovecote that looked like a medieval tower with battlements—it even included slits for shooting arrows at imaginary enemies. A dovecote became a fashionable accessory of country life, even *sans* doves.

In this arrangement, pigeons were not captives. They flew out to find food during the day and returned at night to rest and look after their young. And it's easy to see why they returned. Most large dovecotes were enclosed towers, like cliff faces folded into cylinders, no longer susceptible to bad weather and hungry predators. The inner surfaces were lined with small recesses perfectly suited to pigeon families, often with a round entrance leading to a little rectangular or L-shaped room with perches running along the wall below. There were no uneven, shallow ledges where eggs could slip off. It was a comfortable home. Not all pigeons had such formal housing, however. Often a few pigeons could be kept in an upstairs attic, which came to be called a loft, or even in indentations made on the sides of buildings.

By staying close to people, pigeons ensured themselves a healthy food supply. They are mostly granivores, meaning they eat grains and seed, so it is their good fortune that people based their societies on the cultivation of cereals. Wherever people were, pigeons found fields of grain nearby, a nicer situation than picking up berries and seeds from scrub bushes on windswept cliffs.

Under the care of humans, pigeons spread throughout the world. Domestication and travel brought them into every continent save Antarctica. Often they were brought for food, but not always: Decorative breeds of pigeons were traded across continents as pets, and later racing pigeons were swapped among breeders in different countries.

The first record of *Columba livia* reaching the shores of North America is in 1606, when French ships landed at Port Royal, Nova Scotia, bearing hens and pigeons, which were immediately threatened by the local population of eagles. The navigator of the ship, Samuel de Champlain, later founded Quebec, and when he built his home there he set up a pigeon house. Pigeons came to the New World to live in dovecotes that served as familiar symbols of an upper-class lifestyle. The governor of Virginia received pigeons as a gift in 1621, as did the Massachusetts governor in 1642. In the next century, dovecotes and pigeon houses cropped up in New Orleans, Illinois, Florida, and Detroit, where a Frenchman complained of Indians killing his birds. Estate by estate, pigeons expanded their territory across the continent.

The paradox of living as a livestock animal is, of course, that the very people you depend on for shelter and food are also your primary predators. But the case of pigeons in North America shows why domestication can paradoxically help them thrive. For the first two hundred years of European settlement, the most common pigeon in people's diets was the native passenger pigeon, *Ectopistes migratorius*. Although passenger pigeons and rock pigeons were part of the same taxonomic order, Columbiformes, their habits were very different. Passenger pigeons, which had slender bodies much like mourning doves, lived in enormous colonies that nested in the large forests blanketing North America. When a group of them settled into a forest the entire area would reverberate with the chatter, and branches would snap under their weight. Unlike rock pigeons, they were mobile, migrating from place to place with the seasons, moving in flocks that were reported to take days to pass over a single spot.

The incredible size of their colonies made passenger pigeons an easy target for hunters. The hunting became large-scale slaughter; whole flocks could be caught in nets or smoked out of trees or shot as they flew over. The kills were sold at market for mere cents per bird. The English settlers could indulge their taste for pigeon shoots, but without having to pay a dovecote owner for some of his pigeons; the sportsmen could simply shoot as many birds as they had ammunition to kill. Passenger pigeons once accounted for 25 to 40 percent of the total bird population in the United States. Yet by 1900, they were extinct.

For passenger pigeons and rock pigeons, encounters with humans brought very different fates. Both served as food for humans, but under very different relationships. Animal husbandry involves creating a sustainable supply of animals to eat or sell. A pigeon breeder only kills what can be sacrificed without depleting his stock, since they are his property and his livelihood. Passenger pigeons belonged to no one, in lands that were not yet well protected by law. Although foresight should argue that killing them off would rob everyone of a wonderful source of food, it was no one person's responsibility to care for their welfare. In an atmosphere of every man for himself, it was easy to drive a game animal to extinction.

Meanwhile, the forests where passenger pigeons lived were being cleared to make way for buildings and agricultural fields. Even as people destroyed habitat of passenger pigeons, they were creating new habitat for rock pigeons.

After the passenger pigeon disappeared, along with much of the other wild game in the country, people finally realized they had to act as collective caretakers of wild animals. The government passed laws protecting wild game and limiting indiscriminate killing. This set the stage for domestic pigeons to take the place of wild birds as the delicacy of choice at fashionable restaurants. The rock pigeon had a niche in the New World.

It has always been this way throughout their history: Pigeons have been a secondary or niche food, a delicacy. Because of that, their evolutionary history is different from that of the chicken, which has taken such a starring role in the modern diet. In domesticating pigeons, people took on a more complicated relationship.

Keeping pigeons certainly had benefits, the first being, of course, their meat. Pigeons are best eaten at about a month in age, when they are called squabs. Squabs don't yet have the neat, round bodies of their parents; they're rougher, with lumpy overgrown beaks, and the undersides of their wings are lined with thin yellow plumage known as pin feathers. A squab is ready for killing between four and five weeks old, when the pin feathers have just filled out and it has begun to look more like an adult pigeon. After that the flesh quickly toughens and the bird loses its plumpness as its muscles develop with movement and flight. Squab is a dark, tender meat with a rich flavor; it has typically been served as a delicacy and fetched a high price.

Another benefit was dung; though such an irritant today on buildings and statues, it was prized as a fertilizer, and sometimes the dung was as much reason to keep pigeons as their meat. Concentrating the birds in dovecote towers allowed the dung to pile up where it could be harvested from time to time. Persian dovecotes, which were often large, ornately decorated towers housing thousands of pigeons, were used primarily to generate fertilizer rather than for food. The dovecotes of Egypt helped the farmers of the Nile valley until chemical fertilizers replaced dung.

Dung also was used to tan leather, and in the seventeenth and eighteenth centuries it provided a source of saltpeter, an ingredient needed to manufacture gunpowder. British dovecote owners were sometimes forced to submit to regular collections by the government. Pigeons received little reward for their contribution: Firearms brought an interest in shooting for sport, and in England pigeons were among the most popular targets. Unlike many other game birds, they maneuvered adroitly in the air and made challenging targets, and some dovecote owners began to raise pigeons specifically to supply targets for these shoots.

Their role as a niche food soon took on symbolic significance. Dovecotes emerged in Europe in a feudal system of agriculture, and served as a particular point of controversy. At that time, lords had the dovecotes; in France, tenant farmers were even forbidden by law to keep pigeons. The pigeons of the upper classes were allowed to forage freely on the grain in the fields, and the sight of a nobleman's birds fattening themselves at the expense of farmers became a point of contention. During the French Revolution peasants destroyed noblemen's pigeon houses as a sign of opposition.

A seventeenth-century rant by an Englishman singles out pigeons as a sore point in that kingdom as well. *The Commons Complaint* criticized the use of land resources in England, particularly the destruction and waste of trees and the lack of sufficient food. To remedy the problem, the author offered four solutions: plant fruit trees for profit, breed chickens, destroy vermin, and get rid of England's dovecotes. He estimated that getting rid of pigeons would save millions of pounds worth of corn per year, and implored pigeon breeders to keep chickens instead.

One might wonder whether chickens really consume fewer resources than pigeons, or whether the sight of pigeons feeding in fields was simply a more visible reminder of their cost. The debate about whether pigeons took more food than they produced continued for centuries. An 1830 book on agriculture includes this judgment: "Pigeons are not very profitable to farmers as they eat a great deal of corn, and do considerable damage to young crops." George Walton, who bred pigeons for show, addressed the matter a few decades later, writing that "it would not be right to conceal our belief . . . that the question of whether pigeons 'pay' when kept merely to supply the larder depends chiefly upon the neighborhoods where they are kept. In other and plain words, they do undoubtedly forage in the fields, and it is only where they can obtain a good proportion of food from the land adjoining, only costing the actual proprietor a handful of grain morning and evening, that they can be said to pay as far as concerns a family meal. Even if you have to provide the food there may be little loss if the price of the parents is taken into account. But it's a high price for 'mere food,' and it can hardly be doubted that the parents must live very largely upon the neighborhood—not to say 'the neighbors'—for the produce to be directly remunerative." In other words, pigeons had to sneak a little grain on the sly in order to profit their owner.

But another breeder, Bernard P. Brent, argued that pigeons actually do farmers a service. He complained that the estimates of yearly destruction were usually made by killing a pigeon, looking inside its crop (a pouch in its esophagus that stores food), and multiplying the amount of grain by 365. Such estimates, he protested, were made at sowing time, when pigeons were mostly eating waste grain anyway, and that at other times of the year they may feed on the seeds of weeds, "in devouring which they render a great service to the farmer." Brent even makes a plea for understanding. "I verily believe that much of this bigotry respecting pigeons arises from ignorance, or is made the plea for having a pie at one's neighbor's expense."

In Europe, a dovecote or pigeon house was usually an adjunct to a farm or just a backyard operation. America never had the extensive system of dovecotes that Europe and the Middle East did, but at the beginning of the twentieth century, breeders began experimenting with commercial-scale squab farming, trying to turn what was usually a sideline into a reliable livelihood with steady profit. At that time, T. G. Johnson's Pigeon Ranch near Los Angeles housed thousands of pigeons, and its success spawned commercial ventures all over the country. With the push to commercialize pigeon-keeping, breeders began to look at improving their birds. They imported specialized breeds from Europe, like the Runt and the Carneau, which were larger than standard dovecote pigeons. A few new breeds emerged, so-called "utility" breeds like

A swarm of pigeons covered T. G. Johnson's Pigeon Ranch near Los Angeles. (*Security Pacific Collection/Los Angeles Public Library*)

the King, Mondain, and Giant Homer. New Jersey became a center of squab production, including a large pigeon plant that featured a breed of broad white pigeons called White Kings.

The high price of squab seemed to guarantee a nice profit margin for relatively little investment. In fact, there was something of a "squab rush" in the early twentieth century, with manuals and advertisements perpetuating the idea that great fortunes were to be made in pigeons. In his "bible" on pigeons, Wendell Levi, himself a successful pigeon breeder and scientist, wrote of the "Circean lure" of the squab business, which trapped many naïve entrepreneurs. "There was a general impression that when one had failed to succeed in other pursuits and had a few thousand dollars to invest, he could buy a hundred pairs of pigeons, toss them a little food each day, see that they had water to drink, and sit back and become rich."

Perhaps in reaction to this optimism, other guidebooks cautioned their readers that profiting from pigeons was indeed hard work. A guide from 1922 by breeder Arthur Hazard advocated keeping pigeons as a hobby and extra source of income. "Too much has been written about the enormous profits to be realized from squab plants," he complained, warning his readers of shady pigeon dealers trying to pawn off bad stock to unsuspecting buyers. "Unscrupulous men have painted glaring pictures of such profits merely to sell the beginner a few birds. This is wrong and it is to the interest of the pigeon fraternity to discourage such schemes." The fact that Hazard thought it important to include such warnings suggests that pigeon breeding did not produce outstanding profits.

When the Great Depression hit, squab prices plummeted to twenty-five cents per bird. This was hard upon squab breeders, but it also brought pigeons out of the realm of an upscale delicacy and more in the range of chicken or beef. Since more people could afford pigeon, the new exposure offered an opportunity to expand beyond a niche market. Because pigeon flesh is low in fat but rich in iron and easily digestible, it was popular fare at hospitals and health resorts.

In 1944, Sheppard Haynes, a breeder and member of the U.S. Department of Agriculture, estimated that pigeon plants in the U.S. were producing over 100,000 squabs per year. At the same time, Haynes noted, "there are many monuments to the squab industry around the country today in the form of abandoned plants." Despite the best efforts of squab raisers, as food production became increasingly industrialized, pigeons could never quite compete with the ubiquitous chicken.

In my library copy of Arthur Hazard's book, I found a telling handwritten draft of a letter to an unnamed "Pigeon Journal Editor" that detailed the unsigned author's success at convincing his local tea room to serve squab instead of chicken. "Tea Rooms today have established themselves as one of our American Institutions. They are here to stay," he opined in careful cursive, and predicted that soon more and more tea rooms would start looking for squab "to replace the immortal Chicken." At the letter's close, he issued a rallying cry: "Go after this business fellow Squab Raisers. It's good business, and a good trade for better prices in the summer months when Squabs are plentiful. There are opportunities such as this one right near you. Go after it. Stop hollering Calamity, start dreaming, talking, thinking, eating Prosperity and you'll see the crowd follow."

Well, no: Today we don't see plastic-wrapped pigeon parts in every grocery store and pigeon nuggets in fast-food joints. Haynes had attributed the abandoned squab plants to a lack of experience and knowledge, but other obstacles have kept squab farms from expanding. Pigeons have a well-deserved reputation as reproductive powerhouses; in their early history they were revered for their fecundity. But while pigeons raise young at a steady pace, they can't compare in productivity to chickens. Game birds, including chickens, turkeys, and geese, are precocial birds; when they hatch from their eggs, they literally hit the ground running, ready to find their own food. A newly hatched chick even carries its first meal in its belly, leftover from its large, nutritious egg, so it can make the trip from a hatchery to a farm without needing any food.

Pigeons are altricial—they are born naked and blind and too weak to walk. They rely on their parents to care for and feed them for nearly a month before they are ready to leave the nest. To feed the babies in their first days, both parents secrete a substance called crop milk, which is similar to the milk of mammals. Afterward, they regurgitate grain into their baby's beaks, and gradually teach them how to eat on their own.

For a farmer, keeping pigeons in dovecotes or lofts as a side venture has some advantages. The pigeon parents can be left to feed and raise their young without the expense of incubating eggs or caring for young chicks. It's a do-it-yourself system that doesn't require much equipment beyond rudimentary housing. But in a large-scale commercial enterprise, waiting for all that parenting is highly inefficient, and costs are always high.

Chickens, however, are perfectly suited to mass production, in which quantities are large and prices are low. A hen can lay more than two hundred eggs a year, while a pair of pigeons can raise about a dozen squabs.

So pigeons were never productive enough to be the center of agriculture, but they were popular enough to be taken as guests wherever people went. It didn't guarantee they would conquer the world, but it certainly set them up for domination. And their status as a "fringe" food might even have been an advantage. Few people depended on pigeons for their livelihood, and so they had no qualms about letting the birds fly freely and trusting them to return each night. The birds weren't penned or corralled or otherwise controlled. Many domestic animals are so transformed through breeding that they become dependent on their owners; with pigeons, this was never the case. The eighteenth-century naturalist Comte de Buffon said it this way:

> [Pigeons] really are not domestics like dogs and horses: or prisoners like fowls: they are rather voluntary captives, transient guests who continue to reside in the dwellings assigned them only because they like it and are pleased with the situation which affords them abundance of food, and all the conveniences and comforts of life.

Today, the largest single pigeon operation in the U.S. is the Palmetto Pigeon Plant in South Carolina, which has managed to stay in business since Wendell Levi helped to open it in the 1920s. Since then, the farm has acquired new land, better buildings, and a few automated systems, but for the most part the pigeons are housed like they were in the early days. Because the pigeons must be free to parent their young, they can't be kept in individual cages. Their stacked rows of nest boxes surround an open area where the birds can feed and move about. Taking care of the pigeons is labor intensive compared to an automated chicken farm; the current owner, Tony Barwick, says his staff must check each nest by hand a couple of times a week.

Squab is still sold as a delicacy, particularly at high-class French and Italian restaurants. But the business that now keeps Barwick's operation afloat is a growing population of Asian and other immigrant groups, who have maintained the tradition of eating squab. The largest market by far for squab is found in the country's Chinatowns, and the plant is also setting up Halal inspection to meet the high demand for squab in the American Muslim population.

Barwick is optimistic about the future of squab because of the changing nature of American demographics and eating patterns. He grew up in a farming family, and says that squab was once a cheap and easy meal for farmers in the South who would keep a few pigeons at the top of the barn and eat them as necessary. As those families became more urban, he said, they lost the taste for squab—and the ability to pay for it themselves. "When I first started here twenty years ago, I couldn't sell squab south of the Mason-Dixon Line. They were too poor," he said. Now, everyone seems to be eating out more and willing to pay more for meals. As the price of restaurant dinners creeps upward, the cost of squab is approaching the range of, say, a nice steak. With the growing taste for luxury cuisine, Barwick's company has made inroads in its home region; it now sells about a thousand squabs a week in the South.

British historian Joan Thirsk included pigeon raising as one of several "alternative" agricultural practices that enjoy periods of popularity. Alternative crops and livestock, she argues, become popular only when there is an excess of farming's staples, cereals. After the Black Death wiped out a third of the population in Europe, for instance, fewer people meant less demand for grain, and dovecotes became common. Another pigeon boom came in the late seventeenth and early eighteenth centuries, when food prices dropped and grain could be spared for pigeons. Now, as Barwick noted, people have more money to spend on food and more exotic tastes, which may again bring a resurgence in squab, at least in the United States. But the squab business today has a new obstacle: the growing sentiment that pigeons are urban vermin, not precious meat.

It took a little searching to locate squab in Boston. I found it on the menu of some restaurants in Chinatown and one or two expensive restaurants. Determined to try some, I opted for a small, swanky place called No9 Park that sits at the top corner of Boston Common, just up the hill from the bench where I had watched flocks of street pigeons feeding on handouts and performing fly-bys over teenage girls. My fiancé and I were seated at a table against a row of tall windows with a view of the park—by now the pigeons had long since returned to their nests to sleep.

Here, the "rat of the sky" sold for thirty-seven dollars. When I ordered the entrée, the young waiter asked me how I liked it cooked and I admitted I'd never had squab before.

"Well, it's a dark meat bird," he said, adding casually, "some people call it *pigeon*."

He used the French pronunciation, so it sounded like "pee-*zhon*," and reminded me of when people joke that they bought their coffee table at Tar-*zhay* instead of Target. The waiter suggested I try it cooked medium, since the rare meat could be intimidating to a first-timer.

As I sipped my wine and waited for the dish to arrive, I felt some trepidation now that I had been studying and observing pigeons for quite some time. Was it wrong for an observer to eat her subject? Or worse, would I balk and waste thirty-seven dollars plus tax and tip? Would it be like high-school biology class, when I couldn't bring myself to dissect a fetal pig?

I told myself that pigeons had spent thousands of years being feasted upon by people. If they hadn't, they never would have made it to Boston Common or most of the other city parks where they now live. Perhaps the only way to fully study a subject like this, I reasoned, was gastronomically.

The dish arrived—it wasn't a whole squab as I feared, but an assortment of small browned legs and thighs atop a mushroom puree, with purple rice, stir-fried wild mushrooms, and some wild native vegetable I'd never heard of, all coated in a soy-based sauce. I took my knife and sectioned out a piece of thigh, exposing the vermillion meat beneath. I nudged a little mushroom puree on it for good measure and brought the bite to my mouth. The meat was rich, not fatty like duck or bland like chicken. My qualms vanished. I now had a better grasp of why people bothered to take these birds across continents and oceans: They were delicious.

3

Darwin's Metaphor

One day in late spring of 1855, Charles Darwin made the sixteen-mile trip from his home in the English countryside into London. The purpose of the trip was to embark on a new project; after spending a week in the city, he planned to buy a few pigeons and take them home with him. Darwin was not interested in common pigeons like the kind that lived in dovecotes all over England. He was looking for fancy pigeons—breeds that for centuries had been shaped into new, unusual forms, similar to breeds of dogs, cats, and other domesticated animals that now bear little resemblance to their wild ancestors. Darwin had little experience with pigeons and no great love for them. His grandfather had kept a few, but the naturalist could not even remember having seen a baby pigeon. The interest he took in them was purely intellectual; he thought the birds might help him in a large endeavor that had been preoccupying him for several years.

Darwin visited a poultry dealer on Fleet Street named John Bailey. He chose two breeds, Fantails and Pouters, and paid twenty shillings a pair. Perhaps he chose these breeds because he was interested in extremes: Both were examples of how un-pigeon-like a pigeon could look, and anyone who saw either without knowing what it was might never guess it was a pigeon. Fantails had been bred to carry large, splayed turkey tails that flared upwards; to support this weight they had monstrously large breasts. Pouters, rather than having bodies that were horizontal from breast to tail like a typical pigeon, stood upright atop long legs. A particularly large esophagus allowed the birds' chests to blow up with air, creating a bulb several inches in diameter. Their unusually upright posture lent itself to caricature; an illustration from the time parodied a Pouter as a haughty man with an overstuffed cravat wearing tuxedo with tails.

To his dismay, Darwin's Pouters plucked the feathers from his Fantails on the journey home. But despite a rough start, Darwin got his new adoptees housed in the aviary he had built for them, an oblong hexagon ten feet across and sixteen feet wide.

And so the eminent naturalist had become a novice pigeon breeder. It was less than a month before that he had first showed an interest in pigeons, when he wrote to his friend William Fox, who kept a "Noah's ark" of animals at his house, asking whether Fox had any pigeons, particularly Fantails. Darwin, who had gained fame for cataloging exotic South American species on his journey on HMS *Beagle*, had lately been plying his friends and acquaintances for information on their chickens, cows, geese, dogs, crops, and flowers. He wanted to know how they varied, and how changes in breeds were inherited.

All of it was in the name of some large-scale project he referred to only vaguely; he told Fox that he was busy at work collecting facts for a book "*for and versus* the immutability of species." Darwin wanted to know when the tail feathers of young Fantail pigeons grew in, so they could be counted. Fantails have different numbers of tail feathers and far more of them than other pigeons. At the time, he seemed focused on simply answering the questions on his mind without getting involved. He wrote to Fox: "I must either breed myself (which is no amusement, but a horrid bore to me) the pigeons or buy them young."

Just a few days later, the ornithologist William Yarrell convinced Darwin that pigeons deserved a closer look. Yarrell had written a large natural history of British birds, and his entry on the Rock Dove gives some indication of why domestic pigeons intrigued him. He was convinced that domestic pigeons, even the fancy breeds, all came from the wild pigeon *Columba livia*. He then devoted some space to describing fancy pigeon breeds, which he believed were "maintained and perpetuated by selection and restriction." What struck Yarrell was the amazing variety one could see in pigeons. In some cases, changes had been made in the feathers, as in Jacobins, whose necks were swathed in luxurious hoods that reach up around their faces. But the changes were not just superficial—the essential structures of the pigeons had been modified. Their beaks differed by more than an inch in length, and their skulls had different shapes and sizes.

Yarrell knew that Darwin was interested in variation, and pigeons offered it in spades. After discussing the matter with Yarrell, Darwin had the pigeon house built and bought his first birds in London.

Despite his initial reluctance, once he got the birds settled, his opinion about them changed. He made contacts with pigeon breeders to learn more about the birds. He expanded his dovecote, buying several other breeds. A letter to Fox just a few months after he grumbled about having to breed pigeons demonstrated the excitement of a budding fancier. He ecstatically listed his recent purchases:

> I have now
>
> Fan-tails
> Pouters
> Runts
> !!! Jacobins !!!
> Barbs
> Dragons
> Swallows
> !!! Almond Tumblers !!!
> !!!

Darwin intended to study the differences between the breeds, the same way naturalists would study the different features of birds in the wild to categorize them. But already he had become attached to them. When he invited Charles Lyell and his wife to visit his home, he added that he would show them his pigeons, "the greatest treat, in my opinion, which can be offered to [a] human being." He told his friend that he would kill the birds, remove and preserve the skeletons, and study their inner anatomy. But in a letter dated just a few days later, Darwin admitted to another friend, "I am getting on with my Pigeon Fancy & now have pairs of nine very distinct varieties, & I love them to that extent that I cannot bear to kill & skeletonise them."

Darwin was smitten with his pigeons, but his interest in them was not just sentimental: Pigeons were to become a crucial, if unlikely, crux in the argument he had been working on for years. He was busy putting together a mass of facts and arguments that would eventually become his celebrated book *On the Origin of Species*, a book that would overturn scientific convention and cause massive change in science, culture, and society.

By Darwin's time, the "pigeon fancy" had become a popular pastime among middle and upper classes. Pigeon fanciers in England formed societies and clubs and gathered together for regular meetings. They wrote extensively about their hobby in books and periodicals, discussing both the practicalities of keeping pigeons and the finer points of aesthetics for each breed. At shows, the birds competed for prizes and drew admiring crowds.

Breeding pigeons for show rather than meat was not new. Many of the breeds in England at the time had a centuries-long history. Their lineage could be traced to ancient civilizations in the Middle East, India, and Asia, with some arising more recently in Italy and Germany. In the first century AD, Pliny mentioned fancy pigeons and said they had been bred for some time. Darwin's Pouters and Fantails, for instance, were described in a Spanish text from 1150, and were among the dozen or so breeds kept by the sixteenth-century Mogul emperor Akbar, who was said to have kept twenty thousand birds. Pigeon breeding fit very nicely into the society of Victorian England, which valued the ability to control and shape nature. Like a perfectly ordered and manicured garden, a fancy pigeon was a testament to human ability and creativity. It was also a hobby that could be widely enjoyed. The birds are relatively low-maintenance, and anyone could keep a few pairs in a small coop in a yard or on a roof.

At that time, Darwin was in his forties. His account of the species he had seen and collected during his travels on HMS *Beagle* had helped cement the young naturalist's reputation, and now he was an eminent figure in his field. But although Darwin had been at work for nearly two decades since the *Beagle* trip, he had not yet published the work we most know him for, the book that introduced natural selection, the process by which species evolve. Those intervening years, which Darwin spent in his home in the small country village of Down, are sometimes forgotten in the quick story of Darwin's accomplishments. The story of the *Beagle* makes the discovery of natural selection sound like an adventure story. But although his travels provided the spark the prompted him to study evolution, it did not give him the evidence to develop an overarching theory.

In those days, "naturalism," as natural science was known, involved collection, observation, and categorization. Explorers like Darwin were expected to gather specimens of plants and animals and to comment on the geology of the places they saw as well as the flora and fauna. In the Galápagos Islands, Darwin observed minute differences between similar species on different islands. He did not fully understand what this phenomenon might mean until several months later. It seemed to him as if these organisms, separated by distance, had diverged from a common origin. He realized then that the border between species, always seen as fixed, may in fact be unstable.

It was not until an ornithologist told Darwin that his specimens of three mockingbirds from three Galápagos islands were separate species, not just varieties of the same species, that Darwin had his first solid evidence. It seemed as if the three mockingbirds had come from the same ancestor but, isolated on three separate islands, had gradually grown apart.

This evidence, though compelling, was not enough to announce a conversion to evolutionism. Naturalism until that point was largely shaped by religious beliefs; species were seen as fixed creations of God. In his days at Cambridge, Darwin had studied his grandfather

Erasmus Darwin's poetry about evolution as well as theories by the French scientist Jean-Baptiste Lamarck. Neither of these had been widely convincing; Lamarck's book had been roundly denounced by the scientific community for lack of evidence to support his theories, while Erasmus Darwin's poems were not taken seriously as science. Darwin spent years scribbling his theories of the "transmutation" of species into his private notebooks before he had the courage to broach the topic even with his close colleagues.

It was not enough simply to believe that evolution happens, or even to express it compellingly. Such a revolutionary idea needed strong evidence, a mass of irrefutable information. Darwin was not a philosopher but a collector and a lover of minutiae, so caution suited his nature. And so, instead of putting together an audacious argument about evolution, he embarked on an eight-year study of barnacles.

Peering at the faintest features of these small creatures from his study at home in the English countryside, Darwin discovered an incredible degree of variability. Barnacles—nearly immobile cindercones that barely seemed alive to a casual observer—possessed not only inexplicable differences in form, but also, on further study, a shocking degree of sexual diversity: Some were hermaphrodites, others had two male organs, and others were females that mated with tiny, parasitic males. One could imagine how the hermaphrodites could have gradually separated into two sexes. Darwin saw that the species could be connected in a kind of family tree, a line of descent representing a gradual divergence in form. Their distinctions seemed less defined, like points in a borderless territory of possible forms.

But evolution was only speculation without a mechanism to show how it might occur. Through his observations of the species in the Galápagos, Darwin began to see how the constant struggle for survival, with the individuals best adapted to their conditions favored to survive, could gradually over long periods of time shape the characteristics of living creatures. But any error or any gap in his knowledge could leave him vulnerable to contempt and criticism. The barnacle study convinced him that close and careful scrutiny of organisms, more than philosophizing, was the path to developing a coherent theory of how species change. He would buffer himself with a fortress of facts, build an impeccable wall of data about variability and inheritance drawn from the world around him.

This search brought Darwin to the realm of domesticated plants and animals. Making the argument of evolution was difficult. Creatures evolved over millions of years, and the proof of these changes was lost, its history invisible to the present. Darwin had specimens from his travels and from other scientific collections, but there were immense gaps in their records. As with the mockingbirds, he had compelling similarities but no proof of direct links between species.

Domesticated animals, on the other hand, had been observed and recorded under the eye of history, and intense breeding had speeded their evolution into an observable time frame. To a scientific world still enthralled with encountering new species and exploring new lands, nature was still very much a thing apart from human society. Domesticated animals were degenerate, unnatural, and not worth the attention of a serious naturalist. Darwin felt that the changes wrought by man on animals were a close parallel to the changes that he believed must occur in nature over long periods of time.

Not long after buying his birds, Darwin wrote to his son William: "I am going up to London this evening & I shall start quite late, for I want to attend a meeting of the Columbarian Society, which meets at 7 o'clock near London Bridge. I think I shall belong to this Soc. where, I fancy, I shall meet a strange set of odd men." Rather than hold himself distant from animal breeders, as many other naturalists might, he plunged himself into it. Mixing with pigeon breeders for Darwin involved stepping out of his elite social circle. Even more than class differences, pigeon fanciers seemed strange because of the peculiar passion with which they regarded their birds. Darwin called the fancier Bernard P. Brent a "very queer little fish," commenting on his lack of social graces. Describing another "odd little man" he remarked to William, "all Pigeon Fanciers are little men I begin to think." But when addressing these pigeon men, he was always a deferential and eager learner.

Though scientists rarely mixed with such hobbyists, Darwin was willing to make the journey to the other side, and even enjoyed it. He could study the birds—and their masters—just as he would some exotic new species. Darwin soon joined two fancier clubs, met with prominent breeders, and attended poultry shows. His interest in pigeons also became a fascination with pigeon fanciers, because it was these men, with their strange preferences and minute discriminations of form, that had produced these astonishingly different breeds.

Pigeon breeding was an aesthetic enterprise. There was always a sense that the living birds were approaching some imagined ideal, and that by breeding selectively—choosing the best birds and discarding the rest—breeders could gradually shape the breeds toward the characteristics they wanted. The goal of all of this was to produce the perfect pigeon. Fancier Edmund Star wrote that although each breed followed a different ideal, "the ideal bird of each, whatever the tendency of the variety, is built upon the lines of harmony and perfect symmetry."

As the hobby became more serious, standards became ever more specific, so that judging was seen less as a matter or personal preference and more as an objective assessment of each bird. Each feature of the breeds was defined: the color, beak and wattle, skull, neck, shoulders, breast, flight and tail feathers, thigh—even details like the space between the eye and the beak, or the thickness of the skin ring, or cere, that encircled the eye.

A few key insights made pigeons especially important for Darwin. One was their sheer diversity. There were more than one hundred different varieties that he could find record of. To a naturalist accustomed to noting the finer features of barnacles, this abundantly visible array of differences was "something astonishing," as he would later say in the *Origin*. If these birds were found in the wild, they would be classified as different species, or even different genera. Their colors, feathers, bone structure, even internal organs could be shockingly varied.

But pigeons' diversity would mean very little were it not for another fact: All of the breeds were thought to derive from a single ancestor, the wild Rock Dove. In addition to the historical evidence, Darwin found that by making crosses between very different breeds, he often wound up with offspring that looked much like the typical gray pigeon.

He argued that such birds were reverting to their ancestral form. Pigeons offered Darwin his best evidence that wildly different forms could spring from a single source.

The principles of genetic inheritance were still unknown at the time. But Darwin saw that while naturalists might debate the idea of inheritance, animal breeders were already actively using these principles to effect great changes. Following the principle "like begets like," they knew that in order to obtain a bird with longer feathers or a shorter beak, they must choose parents that best display those qualities. But they also knew that inheritance was not always so simple—some traits seemed to get "buried" behind other traits, only to surface a generation or two later. Through trial and error, they figured out how to obtain different color combinations. Even though fanciers were not "naturalists," they had built up an incredible body of knowledge about animals. They knew all the details about the life cycle of their birds: how often they reproduced, how they raised their young, what sorts of diseases they were susceptible to. Naturalists might look down on animal breeding as unscientific and outside "Nature," but breeders probably knew much more about the nascent sciences of biology and genetics, if only because they had such better access to their specimens.

"Selection," Darwin wrote, "whether methodical or unconscious, always tending towards an extreme point, together with the neglect and slow extinction of the intermediate and less-valued forms, is the key which unlocks the mystery of how man has produced such wonderful results." This simple process of selection had worked countless times in countless plants and animals.

He could envision how these breeds had bloomed from a single origin, gradually diverging over generations. This image was a pointed contrast to the view of nature envisioned by other naturalists of the time. They saw the diversity of life as if it were a flat snapshot, a two-dimensional array of living things, a canvas painted by God. But in fact the picture had depth, it had the added dimension of time, and one could trace back through time from diversity to unity.

Like the barnacles that Darwin had scrutinized for several years, pigeon breeds could be linked in a tree of descent, which Darwin later attempted to organize in his most complete documentation of variation in a species. With his knowledge of fanciers and his observations of the varieties of pigeons, Darwin could now see the birds in a new way. They were not immutable products of creation. Each breed was a living history, the physical embodiment of a relationship between man and animal that stretched back for millennia. Pigeons were always in flux. If domesticated animals could be altered so easily, surely nature—a far greater power in Darwin's opinion—could do much more.

In nature, no breeder selected the best specimens to reproduce. But there were conditions that led one individual to survive and breed while another one perished. Darwin thought about the simple mathematical fact that while individual animals reproduce abundantly, most populations were stable. Where do all these offspring go? Why does every species not grow exponentially? It could only be explained by the strict control of natural conditions—predation by other animals, a lack of food or other resources, an inability to find a mate and reproduce. So much teeming life meant that failure and death were equally present. With such draconian standards for survival and reproductive success, any small advantage could make an enormous difference. Gradually, over many generations—far more generations than are required by active breeding—the individuals with traits that were more successful prevailed. Creatures that were given enough opportunity to diverge— like the inhabitants of islands that were separated from one another by water—would become so unyoked from one another's evolution that they could form different varieties and even separate species.

Darwin's theory had a vital weakness; it rested on assumptions about natural phenomena that are not easily seen or measured. No one had seen the animals on the Galápagos undergoing changes over generations. No one could prove that God did not simply create barnacles as a series of gradually varying forms. Believing in natural selection required people to connect the dots—to envision a continuously evolving path behind the snapshot. It required imagination.

For all his emphasis on gathering facts to avoid the pitfalls of previous evolutionists like his grandfather Erasmus, when it came time to present his ideas, Darwin found himself in need of a metaphor.

Darwin felt that fancy pigeons fulfilled that need. In domesticated animals he had evidence—thousands of years of documented evidence—of change. If his audience could understand his vision of change by breeding, they could more easily imagine a similar process in nature. And using a harmless gentleman's hobby like pigeon breeding could help ease his audience into a vision of the world that was starkly different, even frightening. Fancy pigeons—coddled, ridiculous, grossly manipulated birds—could serve as a metaphor for the mutability of life.

In April of 1856, Charles Darwin met with his friend and mentor Charles Lyell. Up to this point, Darwin had spent years hunting and hoarding facts on an array of different species, and had been tending and breeding his pigeons for the past year. Finally, he felt ready to broach the shocking implications of his work with a trusted friend. Lyell's *Principles of Geology* had been one of the books Darwin pored over during his *Beagle* voyage, and one of the most important scientific works of the period. It argued that features of the earth could be explained by slow, gradual change over time—millions of years of time, a history vastly longer than the events of the Bible. But Lyell, despite his revolutionary thinking in geology, was opposed to "transmutation" theories that asserted that species could evolve from one another.

Darwin presented his facts to Lyell, and told his friend about his ideas of the fluidity of species. After their meeting, Lyell wrote to Darwin, urging him to "publish some small fragment of your data *pigeons* if you please & so out with the theory & let it take date—& be cited—& understood." Privately, however, he lamented in his journal that the theory "brings man into the same system of progressive evolution on which developed the orang out of an oyster."

The theory that so disturbed Lyell was in many ways familiar, as it echoed some of the ideas of past proponents of evolution. But Darwin was not simply arguing that higher animals had evolved from lower ones, and that species boundaries were not fixed. He also offered a mechanism for how it all happened. It was the mechanism—which he called natural selection—that made the idea of evolution plausible.

And the facts that he had carefully collected over the years provided compelling evidence of selection in action.

The following year, Darwin sent the American botanist Asa Gray a sketch of his description of selection, "the briefest abstract of my notions on the means by which nature makes her species." As always, he couched the offering in the humblest apologies: "Without some reflexion it will appear all rubbish; perhaps it will appear so after reflexion."

This sketch contains much of the same framework and main points that Darwin would later use in his public presentation of the theory and in the *Origin*. Darwin chose to begin his argument with his observations on breeding, including pigeons. When he introduced natural selection, Darwin used the metaphor of a breeder to introduce this new concept. "Now suppose there was a being," he wrote, "who did not judge by mere external appearance, but could study the whole internal organization—who was never capricious,—and who should go on selecting for one end during millions of generations, who will not say what he might not effect!" This "being," was nature herself, as Darwin later explained when describing the workings of natural selection.

Gray was concerned about this anthropomorphizing of nature, a criticism that surprised Darwin, who insisted he meant the term "natural selection" as a way of conveniently "expressing the result of several combined actions." Darwin did not seem overly troubled by the implication of an actor or a mind behind selection. He assured Gray that he would be careful to explain his use of the term, but insisted he must use it, for without the simplifying concept of selection, he would be forced to describe the phenomenon as a formula—a sum of competing forces—an awkward concept that seemed too difficult to grasp.

Eventually, circumstances forced Darwin from carefully guarding his theory indefinitely. When Alfred Russell Wallace, a younger and less established naturalist, wrote to Darwin outlining nearly the same theory of natural selection, Darwin's friends finally forced him to go public with his work. Darwin's letter to Gray, and an earlier sketch he had written, were presented to the Linnaean society along with Wallace's letter, and Darwin soon wrote a longer explication of the theory, which he insisted was only an "abstract" of the immense amount of information he had gathered. This abstract was the *Origin*.

The analogy of artificial selection led the argument, with pigeons center stage. This metaphor turned out to be both a strength and a weakness. It is a testament to the power of Darwin's argument that natural selection became widely accepted even before scientists could offer experimental "proof" of its existence. The analogy to artificial selection undoubtedly helped communicate his theory. But by reaching into the artificial world of breeders to reveal truths about the natural world, Darwin invited criticism from his contemporaries, who saw clear boundaries between artificial and natural processes.

"We all admit the varieties, and the very wide limits of variation, among domestic animals," wrote geologist Adam Sedgwick in 1860 in a critique of the *Origin*. "How very unlike are poodles and greyhounds. Yet they are of one species." Sedgwick was not impressed by Darwin's enthusiasm over the diversity of pigeons; if anything, the idea of nature as a fancier seemed to belittle the power of nature. True, man may alter the forms of domesticated animals. But Darwin had not shown that man, or any earthly force, could make new species. That, Sedgwick, argued, required an act of creation. "But, what do I mean by creation? I reply, the operation of a power quite beyond the powers of a pigeon-fancier, a cross-breeder, or a hybridizer; a power I cannot imitate or comprehend. . . ."

In the wide scope of species that Darwin mentioned in the *Origin*, one was conspicuously absent: man. He stayed mute on that issue until later, but the implication was there; if selection was a universal process, then humans were just as mutable as pigeons. The fixed boundaries between species—and the hierarchies between man and animal—were a fiction.

Though Darwin had freed his vision of the world from the all-seeing eye of a creator, he could not completely free his language. He chose to characterize the effects of all this change as a force. The term "natural selection" implied an active, if imaginary, force at the helm of evolution. In the metaphor of the pigeon fancy, God was usurped by an aesthetic. In her biography of Darwin, historian Janet Browne writes, "When he talked about the 'hidden hand' of selection thereafter, he almost always visualized a pigeon breeder picking a favored bird out of one cage and putting it with another bird, also chosen for favorable attributes. He could not help but anthropomorphise natural selection into a mating ceremony deftly engineered by a wise, all-seeing, and sensible English gentleman."

The idea of evolution as a pigeon breeder made it easier for people to misread his views. Darwin's readers could imagine that evolution in nature meant constant improvement, just as breeders are always seeking to better their stock. In the increasingly industrial and capitalistic world of Victorian England, natural selection seemed to follow the same principles that guided views of capitalism and liberal progress—evolution as an active, inexorable path. It was used to justify a view of society in which individuals competed unfeelingly against each other like pigeons for a prize. Darwin insisted that change happened randomly. Any progress was simply the chance result of the fecundity of nature battling against the harsh conditions of life and death. Only after the *Origin* had been published to both praise and outrage did Darwin admit the problems with his language: He wondered to Lyell if he should have called it "natural preservation."

By 1858, Darwin had fully plumbed the pigeon fancy for its insights; he began to give away his birds and moved his aviary to a remote corner of the yard. The *Origin* and its ideas roiled like a fire across the scientific community and beyond, leaving few fields untouched. Pigeons had played a small but crucial role in Darwin's presentation of the theory of natural selection. They served as a flint spark to give life to the more esoteric concept of natural selection. Later Darwin would carefully catalog his knowledge of pigeons in *Animals and Plants Under Domestication*, where he would cite long passages from the fanciers' literature that so amused him, poking fun at their obsession while clearly admiring it.

Darwin also helped people better understand pigeons by solving a small but important riddle. At the time when he began studying the birds, fanciers and naturalists interpreted the diversity of pigeon breeds differently. Pigeon fanciers firmly believed that their prized varieties represented different classifications and often referred to thc different breeds as "species." But naturalists believed that all the varieties came from the same species, *Columba livia*, and they simply overlooked the changes in form as unnatural aberrations. Few naturalists showed Darwin's interest in the evolution of domesticated species. Darwin's insight was in recognizing both the unity of pigeons and their diversity. Only by bringing together these two views of pigeons could their true story be told.

4

Hopeful Monsters

The first time I read *Origin of Species*, I was befuddled, even disappointed, to find pigeons at the opening of the most famous book in biology, the book that contains what has been called the greatest idea ever conceived. What business had pigeons headlining such a noble book? It seemed even stranger that these birds, which I associated with homogeneity and sameness, were being used as models of diversity and variability.

Like most people, I had never heard of fancy pigeons.

It is difficult to describe how much these birds deviate from the iconic form we associate with "pigeon." When I began to attend pigeon shows around New England, I found myself imagining the crude instructions for how to transform an average street pigeon into each breed.

- Pouter: Squeeze the pigeon from its lower half like toothpaste, until its body is skinny and it breast puffs up. Then pull its legs until they are twice as long as normal pigeon legs. Stand upright.
- Capuchin: Drag its neck feathers out into a saucy boa.

- Jacobin: Take a Capuchin and continue pulling feathers until its head is obscured by a giant saucer-like pompom.
- Modena: Blow the pigeon up into a round bulb until its cheeks begin to bulge.
- Fantail: Affix a large turkey tail to the back of a pigeon. Because the tail will be too heavy, give the bird an enormous breast to compensate and move its head onto its back. Good enough.

Of course, pigeons must be coaxed into these shapes through centuries of careful breeding. Fancy pigeons today weigh from four ounces to four pounds. They come in an array of shapes and an even greater variety of colors and patterns, and feature a range of decorative touches, such as feathered frills and brain-like skin growths on the beaks and around the eyes. Some pigeons have unique calls: The Laughers laugh and the Trumpeters gurgle. Some breeds have neurological tics that cause them to shiver. The Zitterhal has an elongated neck that it "zitters," or pumps up and down—the bird uses the behavior as a courtship ritual. Swallows and Magpies are bred to look more like their namesake birds than like pigeons. Yet each of these breeds is considered *Columba livia*, the same species that populates our streets.

A visit to a large pigeon show, like the Grand Nationals held each year by the National Pigeon Association in the United States, is enough to convince anyone that pigeons must be one of the most malleable creatures on earth. Darwin asserted that there were 150 known varieties. Today there are more than seven hundred, and many more exist that have never made it into the official lists.

When I first became interested in tracking the history of street pigeons, fancy pigeons seemed like an irrelevant offshoot. But Darwin taught me that these worlds that seem separate are not necessarily so. In the elaborate family tree of *Columba livia* that he envisioned, all pigeons were connected by a common ancestor, no matter how different they seemed. The family tree became a story about how the same birds could be transformed through different conditions of life. A pigeon's relationship with people is the condition that most guides its fate. Pigeon fancying represents the most dependent relationship; in exchange for food and shelter, the pigeons give complete genetic control to their owners. In the century and a half since Darwin so carefully inventoried the fancy, much had happened to his family.

This year, the national show took place in a large trade building at the Des Moines State Fairgrounds in Iowa. The first morning of the show was frigid, with temperatures somewhere in the teens, but inside several hundred pigeons stood in the safety of their cages, surrounded by sawdust and dung and scattered seeds and corn. Some had been living here for a day or more, while other birds were still arriving. They came in crates of metal and wood, stacked into vans or loaded onto airplanes, or shipped in large cardboard crates dotted with holes. At about ten in the morning, humans were still relatively scarce, just a few breeders milling about the rows.

The pigeons were organized by breed, and colored signs at the head of each row announced each one: Modena, Old German Owl, Frillback, Fantail, Indian Fantail, Jacobin, Runt, King, and so on. Some of the names were the same as those Darwin listed, others were entirely new. Each pigeon had its own cage, and the birds stood silently, but when I walked past them I could hear feet scuffling as they stood at attention and shifted around the cages.

I met a man named Ron Nelson holding a bird in a row marked "French Mondain." "Here," he said with a friendly drawl, "do you know how to hold a pigeon?" He had long gray hair, a trimmed white mustache that trailed up to his sideburns, large thick glasses, and a trucker hat with a pigeon club logo on it. He took a reddish-brown bird out of its cage, and showed me how to wrap the fingers of one hand firmly against the bird's side and rump, and the other over its breast. The pigeon was blasé; it was clearly used to frequent handling. "That's a big bird," he said, "it weighs about two and a half pounds." Many of the other breeds are all feather, he told me, but French Mondains are a utility breed: They were originally raised for their meat and only later became show birds in their own right. After the breed was recognized by the National Pigeon Association in 1928, Nelson told me, people began breeding it for "show points." I asked him what he meant

A Capuchin strikes a pose.

by that, and he took me to see birds of another breed he keeps—Flying Rollers.

Those birds looked much like street pigeons, just smaller. When Flying Rollers fly, they "fall" from the air in tight, continuous spins like a whirring pinwheel. The only way to judge the performance of flying birds is in the air, which just isn't possible at shows (the exception is Parlor Rollers, which will somersault along the ground for several hundred feet and can be judged on a lawn or gymnasium floor). Without flight as a basis for judging, breeders started to focus on looks—trying to make the birds more pleasing to a judge's eye. "In the 1950s," Nelson said, "a guy named Bill Pinson, who raised Rollers, said that if people kept breeding for show points, someday there will be separate breeds. Eventually in the '80s there was a split." He took me to another row of cages, where the Show Rollers were kept.

"This is what happens when you breed a bird for show points," he said. To an untrained eye, Show Rollers didn't look radically different from the Flying Rollers. But as I looked closely I saw they were bigger, with rounder, fluffier heads. Fluffy feathers are bad for flight, Nelson explained to me, but they look better. "In the last thirty to forty years, these birds have evolved into a separate breed of pigeon." Same pigeons, different aesthetics—one bred to impress with flight, the other to impress with looks. And now they were different breeds.

"Believe me, evolution is not a theory, it's a proven fact right here," Nelson said, nodding his head to indicate the rows of cages all around us. "This is evolution in high gear."

Darwin recognized that evolution occurs under domestication just as it does in the wild. The difference is the conditions that guide it. In the wild, the gene pool of a species can be affected by predators, food availability, climate, loss or gain of habitat. In fancy pigeons, however, evolution is driven by human aesthetics.

In nature, aesthetics can also be a powerful evolutionary force: It's called sexual selection. Passing along one's genes depends on more than just survival—it depends on finding a mate. If brighter plumage or a showier tail helps a male bird get the attention of females, it may make him a more successful breeder even though he is more visible to predators. Darwin alluded to the similarities between breeding and sexual selection, but later scientists have drawn explicit analogies between the two. Pigeon fanciers choose for outward appearance and beauty, and so they cultivate the same sorts of features in their birds that would appeal to potential mates in the wild. And since breeders do, in fact, choose which birds mate together, they can be seen as artificial matchmakers, imposing their own preferences on the mate choices of their pigeons.

With no predators or other stresses putting these flourishes in check, domesticated animals can develop far more outlandishly than they do in nature. No wild bird would survive with its head encased in the giant feathered pompom of a Jacobin. And no pigeon would ever grow so large as the Runt—such size would take too much energy to support and make it impossible to dodge a hawk in flight.

Today's domesticated pigeons fall into three categories: those bred for show, those bred for meat, and those bred for performance, like Flying Tumblers and Racing Homers. You can compare fancy pigeons to dogs, cats, or horses bred for show. But while a show dog may be bred to be beautiful, it also must be trained; its nature must be nurtured. But aside from the breeds that must perform, pigeons are judged almost entirely on their looks. "The pigeon fancier is the artist among breeders," wrote fancier Edward Star in the late 1800s. "His work of living pictures is the outcome, and to satisfy the same longing that incites the painter, the sculptor, or the connoisseur."

But as in art, tastes can change. In England in the eighteenth and nineteenth centuries, the height of fashion was a bird called the Almond Tumbler. Tumblers are among the oldest classes of pigeons, carried to Europe and England from the Middle East. They were first bred for a peculiar pattern of flight. The Roman emperor Frederick wrote of a Syrian pigeon that somersaulted in the air, and a writer in the seventeenth century describes them as throwing themselves backward, tumbling like footballs. They were eventually bred into different varieties, such as Nelson's Rollers. But in eighteenth-century England, fanciers were more concerned with making Tumblers look good for show. They created a variety of Tumbler that was an upright bird with a small round head bulging over an extremely short beak— the English Short-Faced Tumbler. In the process, these birds had lost entirely their ability to fly, and with their short beaks could not even feed their own young; fanciers had to bring in other kinds of pigeons as foster parents.

What distinguished an Almond Tumbler from other varieties of Tumblers was its color. Pigeons carry only three possible color pigments in their feathers, but many different genes can change the way the pigments get distributed on feathers, resulting in a wide variety of colors and patterns. Birds that are "almond" are a mélange of three colors. The body or "ground" color is called yellow—actually a beautiful light chestnut—spangled with flecks of white and black. In a good Almond Tumbler, every wing and tail feather should carry all three colors, clearly separated from one another. The effect is a bird painted with impressionist brushstrokes. Creating this look is extremely time-consuming—it requires grandparents carrying different traits to achieve the right mix.

Perhaps because of the challenge of breeding it, the Almond Tumbler caught the imagination of fanciers in England. They devoted entire treatises to it. "If it was possible for noblemen and gentlemen to know the amazing amount of solace and pleasure derived from the Almond Tumbler, when they begin to understand their properties," wrote John M. Eaton in one such tome in 1851, "I should think that scarce any nobleman or gentleman would be without their aviaries of Almond Tumblers." In addition to their color, Almond Tumblers were noted for their shape—their bodies in profile drawing an elegant S-curve from their legs to the top of their heads. Eaton, captivated by the beauty of the bird, saw aesthetic perfection in its curves. "I am not aware that there is anything under the Sun, or that you can imagine or conceive, that is so truly beautiful and elegant in its proportion or symmetry of style, as the shape or carriage of the Almond Tumbler approaching perfection, in this property, (save lovely woman)," he wrote, his nod to the fairer sex of his species a bit of an afterthought. Darwin, too, was not immune to the birds' charms. When he listed his newly acquired breeds in a letter to a friend, his Almond Tumblers were bracketed by three sets of exclamation points.

Today, the Almond Tumbler is all but forgotten. If they fall out of favor, breeds can go extinct, and new ones are formed all the time. Sometimes a breed is formed from the pressures of competition, like the show birds that evolved from Tumblers and Rollers. A fancier might add a new flourish to create a new breed. For instance, some pigeons come with plain legs and some have a spray of feathers around their feet, called muffs. I met a man at the national show who has spent the last fifteen years putting muffs on Chinese Owls, a muff-less breed of pigeon that has a fluffy cravat of feathers running in reverse direction up their breasts. Introducing the genes for leg muffs is no simple task. The Chinese Owls must be crossed with a muffed breed, which yields short-muffed pigeons that look something between the two. Then the breeder must recultivate the features of Chinese Owls while also bringing a long-muffed breed into the mix to improve the muffs. It takes many generations to create a new line.

As the morning progressed, more people filled the large showroom of the Grand Nationals. The judging began, and the room broke into little fiefs for each breed, surrounded by however many fanciers a breed could attract.

One of the largest crowds formed in the corner of the room where the Modenas, named for the town in Italy where they originated, were being judged. Like Tumblers and Rollers, Modenas were originally flying birds, but eventually they were bred for show, and today are so stout that it's a surprise they can fly at all. They are one of the most popular breeds of pigeon, and there are hundreds of them at shows like this; judging Modenas takes many hours. The birds are shuttled out of their cages into a row of judging cages along a wall, and then taken away as they are eliminated.

As I watched the Modenas, I chatted with Bob Haas, an eighty-year-old man from Pennsylvania who has been breeding Modenas ever since his son brought one home forty years ago. Bob and his son Richard still keep a loft together and take their birds to shows on the East Coast, and Richard regularly judges shows throughout the country. I met Bob at a show in Rheinbeck, New York. Though he walks slowly and his hands are bent with arthritis, he is sharp and resilient. To get to this show, he packed his frail body into a rental van with a half-dozen other, younger people and endured a sixteen-hour drive to Des Moines. When he arrived that morning, he shook his head and told me, "I didn't sleep at all. They had the rap music on and, oh, it was awful."

"This is a beauty pageant," he explained. Features of each breed are divided and described and allotted points, adding up to one hundred. Different features are weighted differently for each breed: For some, the feathers are most important, for others, the overall body shape is key. A few breeds have to perform; Pouters, for instance, must be able to blow up their enormous chests in response to a judge's mating call. Each standard comes with an image of the "ideal bird" of that breed. Tellingly, this image is not a photograph. After all, á true ideal, like one of Plato's forms, does not exist in the real world. Instead there is a

A Modena, "the bird of curves."

simple line drawing of the bird, sometimes from different angles. The drawings, most of which are executed by an artist named Diane Jacky, are smooth, slightly exaggerated, almost childlike. Some clubs will place drawings of the ideal bird directly behind the cages of the real ones, so that everyone knows what the real birds are up against.

At previous shows, Bob had told me about what makes a good Modena. The birds have round upright bodies that, in profile, should be perfectly balanced over their legs like a large, full wineglass. The tail points straight up from the bulbous body, and the head tucks in against a short neck. Though not elegant birds, Modenas have a certain robust shapeliness, as though Rubens had painted a pigeon. The National Modena Club's motto, in fact, is "Promoting the Bird of Curves." Modena breeders are always making the birds bigger and rounder; their heads and faces have become so full that their brows are beginning to bulge over their eyes. If the trend continues, Haas told me, their eyes might get obscured by their brows and cheeks, making it difficult for them to see well enough to eat. Beauty has its trade-offs.

In the double-row of metal cages along one wall, about thirty pure white Modenas stood in cages. Facing them on a row of folding chairs sat an equal number of people, watching. In between was Ron Zittritsch, a silver-haired man in a white lab coat, decorated with patches, that serves as a judge's uniform. He moved slowly from cage to cage, lifting the metal doors, pulling birds out, holding them, turning them, and placing them back with a gentle tug on the tail, the signal for them to stand at attention. The Modenas straightened up obligingly, hopping back and forth a little on their feet. The birds could have been mass-produced—they looked exactly alike to me. All were downy white, thirty voluptuous bird pillows. Some might have been bigger than others, some a little less upright. But for the most part, they seemed about as close to the ink drawing of the perfect Modena as anyone could expect a living bird to be.

I talked with Bob as we watched the judge handle each pigeon-pillow in turn. Bob liked to tell stories about the old days of pigeon fancying: He told me about the time he and his son judged an entire show themselves, about the banquets and live bands at the show he attended in Bermuda, and about the shows he himself had organized decades ago. The world that Bob described seemed far more glamorous than today's event.

"A lot of the breeders here, they're old," Bob remarked slowly. "Look around, they've got white hair, glasses. Some of them have canes; some of them are rolling around!"

It was true; a couple of the prominent breeders here were cruising the hall in padded electric chairs. You go to pigeon shows these days and you see rooms of men well past middle age, and hear more about regulars who have died than newcomers. Pigeon fancying is a fading hobby in the United States. It flourished here in a culture that was more rural, where keeping pigeons was seen as a good way to teach boys responsibility. Now those boys have grown up. A smattering of new competitors do join—usually sons and daughters of other fanciers, or wives who have taken on their husbands' hobby. With fanciers failing to produce a new generation, the hobby is veering toward extinction, like a breed that has become so specialized it has forgotten how to reproduce itself. All sorts of pressures speed its decline. Today, society regards a person who keeps more than two or three animals as suspect. Many pigeon fanciers face zoning restrictions and complaints from neighbors if they are public about their hobby. Pigeons especially have gained the reputation for being dirty and disease-infested, even if they've never lived on the streets. And fear of bird flu outbreaks has made it harder to transport the birds; flu scares have shut down pigeon shows in England.

But when I ask these men why interest in their hobby has declined, one answer prevails: video games.

Young people today, they say, are used to sitting in front of TVs and computers; they are used to being entertained. They don't want to take on a hobby that requires daily effort and discipline. They want instant gratification, not long-term reward. Pigeon fanciers recognize that the world is changing, but they don't want to be a part of the new environment they see.

Eventually the judge began eliminating, handing off white Modenas to stewards who shuttled them back to their cages. The number of pigeons dwindled, and soon he had picked a winner. But these thirty birds were only the first group. Each color was divided by sex—hens and cocks—and age—younger or older than one year. This round was just the young hens. Next it was the young cocks, the old hens, and the old cocks. And that was just the white birds. Most Modenas are exceptionally colorful. They come in two basic categories: Gazzis, which have a white body with colored markings on head, tail, and wings; and Schiettis, which carry their color throughout their body. They are further grouped into patterns: barred, tri-marked, and T-patterned. There are also Argents, which are Gazzis with delicate white lacing throughout the wing. Each pattern can be found in different colors: blue, white, black, red, brown, silver, ocher, khaki, gold, lavender, sulfur, dun, russet, and more.

The judging of the Modenas continued until early evening and then picked up again the next morning. The judges progressed quickly—too quickly, one man murmured to me—and just when we expected the competition to last until the closing of the show, as it often does, it was down to the last competitors. After choosing the best of young and old, hen and cock, and colors, it was time to choose one Modena out of the champion Gazzi and Schietti. Into the cages went the contestants. This day it was a bronze Gazzi, with a blackish head and dark brown wings on a snowy white body, and an all-black Schietti.

The Gazzi judge and Schietti judge conferred for a while as the two birds stood in adjacent cages. The audience was buzzing. As much as a pigeon's qualities are quantified in standards, when the competitors are this closely matched in quality, it's tough to find a clear winner. At this level, discerning the difference between pigeons is quite difficult and depends on the particular likes and dislikes of the judges. One fancier had already told me he believed the black one was the best but, based on the judges, the bronze would win. And he was right. The judges announced the winner, and a whoop went up in the crowd. The black Schietti was shuttled away and the winning breeder took his place next to his bird for photos.

Later, I looked at the winners, back in their cages, surrounded by other, apparently lesser Modenas. Even with help from Bob and the other breeders, I saw that I had no hope of developing the eye for detail necessary to distinguish these birds. Though the old treatises on pigeons may seem laughable with their talk of art and aesthetic, as Darwin said: "We may smile at the solemnity of these precepts but he who laughs will win no prizes."

With all of this focus on form, the pigeons begin to seem like crafts projects, moving statues on display. It's easy to forget these are living birds, carrying out their lives in extraordinary circumstances. They are certainly the most pampered of all pigeons. They receive housing, food, and medicine. No one even expects them to fly hundreds of miles on an empty stomach, as racing pigeons do. Fancy pigeons may live in basements and garages, simple sheds with chicken wire fly-pens, or elaborate heated lofts that look like small houses. They are given food and, perhaps, allowed to fly around an enclosed area. Most are not released into the open air unless they are flying birds—Tumblers, Homers, Rollers, and the like. Their main purpose is to breed. Because pigeons can breed multiple times in a year, each one is more or less expendable; a pigeon fancier thinks of his stock as a genetic pool that is slowly shaped and improved.

The breeding enterprise depends completely on the breeder controlling who mates with whom. And so these birds cannot carry out their native instincts; they are placed with certain birds the breeder deems a good match, and can't remain with a single mate for life as they do in nature. It is inefficient to let a good bird stay with the same partner; a breeder wants to ensure that the best birds have as many mates as possible.

Some pigeons are content to mate with whatever is offered them, but occasionally a pair will not perform. Fanciers have developed methods for coaxing reluctant birds. John Moore's treatise from the 1700s acknowledged that it was difficult to make pigeons "pair to one's mind," and advocated putting two birds in adjacent coops where they could see one another, and be allowed to eat and drink from the same vessels, "feeding them often with hemp feed, which makes them salacious." Privacy seems to be important; many fanciers told me that they essentially follow Moore's advice by keeping pairs in a coop together, or simply putting them behind a divider, which does the trick.

At the shows, you are only seeing a fraction of the pool of pigeons sitting at home in each breeder's loft. Selection doesn't work unless there is a group of individuals that loses out. In breeding, the flipside to selection is culling, the weeding out of undesirable elements from the gene pool. Culling the outcasts is just as necessary to the project of breeding as is selecting the best individuals to breed.

At an earlier pigeon show, I asked a prominent breeder in New England what happened to the birds he didn't like. He gave me a vague answer and seemed to get busier and busier. At the next show

I attended, which he organized, he pulled me aside, told me I was becoming a nuisance, and ordered me to put my notebook away. After some confusion I realized that he thought I was an animal rights activist in disguise.

In Wendell Levi's bible on pigeon breeding, which was written in the mid-1950s before the days of PETA, he describes culling in one blunt paragraph. Whether a breeder is producing birds for food, flying, or fancy, "old birds should be culled religiously," he wrote. "Any bird which is deformed or is noticeably off-standard should be removed from the loft and killed, but never given away or sold." The same fate awaits birds that are sick or "broken-down breeders" that are not producing young.

It's hard to see culling pigeons as a crime, when pigeons have served as livestock animals for thousands of years. Though fanciers are hesitant to talk about their culling practices, some never kill a bird unless it is sick or injured, while others breed so many birds they must regularly kill the ones they don't like. Most fanciers I talked to insisted that they can swap most of the birds they don't want, since they might be of use to other breeders. But only a few people I talked to see their birds as pets to be cared for to the day they die. Pigeon fanciers readily admit love for their birds, but it is not the sentimental love of a pet owner.

Since Darwin, the scientific literature contains scattered papers about feather color in fancy pigeons and the number of the feathers on Fantails, but little else. Despite the range and complexity of variation in fancy pigeons, few modern scientists have taken them up as subjects. It's as if the birds, once they served their purpose in Darwin's arguments, lost their meaning for science. But that might change. Jill Helms, a cell biologist at Stanford University, is championing the idea that fancy pigeons could be an important scientific tool.

Helms studies the question of how different species acquire unique facial features. If you look at early embryos from a wide range of species, their faces look alike, but at some point in their development they diverge: A bird looks like a bird, an elephant looks like an elephant. She is hoping to identify the genes that control facial development and understand how they work. She believes that studying these questions in animals would reveal genes that might be involved in facial malformations in humans.

"To study that kind of question," Helms said, "you have to have an experimental model that exhibits dramatic variation in facial features." So far Helms has studied facial differences between species—a duck and a quail, for instance. But she said that comparing different species is tricky because it's difficult to isolate those genes that account for facial differences from all the other genes that vary between species. Ideally, she said, one would study a single species that displays a large range of variation in shape. At first, Helms thought of domestic dogs, but doing experiments on dog embryos would be slow, onerous, and would raise ethical concerns. Bird embryos, however, develop outside their mothers' body in eggs, and birds breed quickly and can be managed in a laboratory much more easily. "Thinking about it a little further, I came upon the idea of domestic pigeons," she said.

As products of genetic tinkering, pigeons offer an enormous range of variation. And although they belong to a single species and can interbreed, each breed has been maintained for many generations and is genetically distinct from the others. Fertilized eggs are easy to obtain from breeders, and can be studied and manipulated in the lab. As a research subject, Helms said, "the pigeon, I think, rules." In 2007, she published a paper laying out her arguments for using fancy pigeons as a research model, and has been trying to generate interest in a project to sequence the pigeon genome.

John Fondon, an evolutionary geneticist at University of Texas Southwestern Medical Center, also rediscovered Darwin's birds as a potential model for his research on evolution. Darwin and subsequent evolutionary biologists have done a good job of explaining the processes of selection. But much less is known about what produces variation for selection to act upon: What are the tools that allow genetic information to change over time? To answer this, Fondon looks at species that show a high degree of "evolvability" in their shape. "Some species seem to possess a nearly limitless ability to evolve extremely rapidly, while others remain stubbornly unchanged over many millions of years," he explained. For instance, domesticated dogs have been bred into many diverse sizes and shapes over a relatively short period of time—they can be considered more evolutionarily pliable than cats, which do not yield such differences.

Fondon began identifying genetic differences between different-looking dog breeds that might reveal how these differences arise, comparing short-nosed and long-nosed dogs for instance. But to prove he had found a gene that controls length of the nose, he would have to do some old-fashioned genetic crosses. These are the kinds of things you see in textbooks about Mendelian genetics; when two parents with different traits breed, you can learn something about how the trait behaves by looking at subsequent generations.

But, like Helms, he felt that manipulating dogs in this way would be impractical and unethical. "Dealing with this difficulty of not being able to prove anything in dogs is what made me think of Darwin's experiments in pigeons," Fondon said. He had never seen fancy pigeons, but he got hold of a copy of a book by a photographer, Steven Green-Armytage, called *Extraordinary Pigeons*. The book showcases the unique differences among pigeon breeds in elegant portraits. "As someone who studies the evolution of morphology, when you flip though that book, you see pigeon breeds with just outlandish differences in morphology and behavior. The only thing that comes close to it in the animal world is dogs. And even then, if you look at range of variation, I think pigeons win."

Like Darwin, Fondon began to get involved with pigeon fanciers. He quickly discovered the genetics buffs and attended a national pigeon show with them. He came back with a few different birds, and started breeding them in a shed outside his home in Texas. Fondon wasn't interested in doing selective breeding. Instead, he wanted to breed two wildly different breeds together and see what happened to their traits in their offspring. This kind of thing is sacrilege to fanciers, who value purity of their lines and only cross very similar breeds together occasionally to achieve certain features. Indeed, the breeders who sold him birds made him swear never to let these monstrous offspring get into the general breeding pool.

One of the major questions Fondon is trying to answer is whether noticeable differences in shape are caused by a limited number of genes—like the genes that give pigeons their color—or whether they are controlled by a large number of genes that work together. Color genes follow classic rules of Mendelian genetics: one gene, one color feature. By crossing pigeons with different shape characteristics, Fondon can see whether these traits also follow Mendelian rules. If they don't, it poses a hurdle for his research. "We want to find the mutations that are causing the differences among breeds; if it's hundreds of genes, it's an intractable problem," he said.

Fondon has looked into Darwin's writings for clues, since Darwin had also performed crosses on different breeds of pigeons. Darwin was trying to determine whether subsequent generations were always a blend of their ancestors' traits, or if different traits reappeared distinctly. If blending was the rule, then organisms would never diverge; they would just keep merging like different colors of paint becoming a uniform shade. Darwin's theory depended on the tendency of species to vary, because variation is what selection acts on. By crossing pigeons and seeing that their grandchildren inherited a distinct trait belonging to one of their grandparents, not blended trait from both, Darwin was able to dismiss this counterargument to his theory.

But his account of what came of the crosses was vague. Fondon looked for unpublished accounts, which led him on a trip to England to visit Darwin's home and archives at Down and Cambridge. He pored through notebooks and scraps of paper related to the pigeon experiments. Darwin had left a voluminous pile of notes behind, all in atrocious handwriting, and much of it still unsorted. Fondon would find something relevant, and bring it to one of the scholars at the archive, who could usually translate 80 percent of what was written. He uncovered tidbits like: "Blue Roller killed by cat," beneath a date. Darwin reused paper, so sometimes a note would be scribbled on the back of a drawing from one of his children. All of it yielded very little description of his pigeons.

For now, Fondon has had some success crossing a Roller with a Scandaroon, an old rare breed of pigeon with a heavy beak that hooks downward. He is waiting to gather enough grandchildren of this couple to know whether their beak shapes will follow Mendelian rules. He plans to take his garage operation into the laboratory for a full-fledged research project, and ultimately to look at the genetics of behavioral traits as well as outward appearance. If he and Helms have their way, the strange living laboratory that breeders have created could once again come under the scientific lens.

In one of the rows of cages at the Grand Nationals, I happened upon a treasure that I had hoped to find but never expected to. In a small section marked English Short-Faced Tumbler, a tall man with a bald head covered with a baseball cap was holding a small reddish-colored pigeon. He introduced himself as Dan Revolinsky, and the bird, he said, was an Almond Tumbler.

"This is what you see in Fulton's book," he added, referring to the classic text from the 1800s filled with colored plates illustrating different pigeon breeds.

This bird was one of three Almond Tumblers in a row, all his. "They can't preen themselves," he said, indicating the small, pink beak on the pigeon he was holding. He pulled out an emery board and made a few quick strokes along the beak's tip, which he explained had a tendency to grow into a hook. Then he brought out a toothpick and began cleaning the bird's tiny nostrils. "It's not the most glorious job in the world," he noted.

He put the bird back in the cage. It was, I thought, quite pretty. Dainty, with chestnut feathers flecked in black and white. Each of the three birds in front of us was a little different—one a bit dark, one with larger blotches of color. The Almond coloring changes over the bird's lifetime; it is only in its second or third year that the color reaches its peak.

Revolinsky lives in Texas, where he is on active duty in the Coast Guard. His job involves doing things like jumping out of helicopters into the ocean to rescue people, things that most of us witness only in movies. Pigeons are a haven for him, and his wife and four daughters help him take care of the birds despite his erratic schedule. He plans on retiring soon and moving to rural Wisconsin, where he can expand his pigeon operation. Though he keeps a few different breeds now, his goal is to raise the Almond Tumblers.

When Revolinsky brought his finger to the male bird's cage, it rose to attention as it was trained to. It stood in profile to us, so we could see its proud chest curve into that iconic S that I had seen in the old drawings.

"See how he just stood, and his feet came up and his head threw back? There's an imaginary line from his leg all the way up through his eyeball. It should be a perfect line coming up. Wings below the tail, rump slightly raised, tail slightly out—we call it splayed—and the feathers should all be in place." His golden eye fixed us. Even in a sawdust-covered cage smattered with poop, the bird looked regal. I could see why fanciers once raved over its silhouette, even if it lacked the outlandishness of some other breeds. Revolinsky regarded the male bird and the hen next to it. "Those are both, out of a hundred points possible, those birds are both well up in the nineties."

He took one of his other birds out, and opened up its tiny pink beak. He gave it a few quick strokes of his emery board, braced its head lightly against his stomach, and began pulling dust and debris from its tail feathers.

Maintaining a line of Almond birds takes careful planning. Revolinsky runs his loft like a control center, keeping track of every bird and every egg. When he mixes birds together to get Almond, he needs to look two or three generations behind each bird; he must know, for instance, if a pigeon is carrying a recessive color trait, which will emerge again somewhere down the line. "It's like a palette—it's like making a painting," he said. "You've got to put the foundation down and then you've got to start putting all the other colors down."

The Almond pattern comes from a gene that lightens the color of a bird and causes flecking or "break" in the feathers. This gene is found in many pigeon breeds, but in the traditional Almond Tumblers, the Almond gene is mixed with other genes in a particular way to achieve the tri-colored speckles. Almond is a sex-linked trait, meaning that the Almond gene lies on the chromosome that determines the bird's sex. Pigeons don't have "X" and "Y" chromosomes like people do; instead, there is one sex chromosome—males have two copies, females one. So a male can carry two different sex-linked traits, including a recessive trait that is hidden, while "a hen is what she is," as Revolinsky put it.

Once you get the Almonds, you can't simply keep breeding Almonds together, since the gene responsible for the Almond pattern can cause severe health problems or death in males if they get two copies of the gene on both of their sex chromosomes. So if making Almonds is like painting, it's a painting that changes with each generation, and requires constant adjustment through breeding to maintain the right balance of color traits. Revolinsky's pigeons must be kept and bred in individual pens, lest an interloper jump on one of his hens and ruin the careful genetic planning that goes into maintaining the Almond line. The youngsters require foster parents since Short-Faced Tumblers can't feed their young. He and his daughters often hand-feed the babies using syringes and pipettes.

As we talked, a white-haired man in a white coat wheeled a cart into our aisle. "Can I interest you in any prints?" he asked, pulling a few cellophane-wrapped sheets from a stack. "These are not reproductions. They're originals from 1878. From Fulton's book on pigeons." We took the stack and flipped through the brightly colored drawings.

"See, there's the plates," said Revolinsky. "Those are flying Tumblers and those are the English." In each illustration, three or more birds were laid out in profile against a background of a loft. We found two prints of Almond Tumblers, and Revolinsky pointed out the different variations on the coloring in birds of different ages. "Thank you, I already have them," he said to the man, handing the prints back.

"Everybody thinks they're neat," he said as we turned back to the birds. "They all admire this but none of them will really take on the task. It's sad because it's really the heritage of these fancy pigeons." Even the judges here may have never seen an Almond Tumbler—they can only judge them against the sheet of paper holding the standard, which Revolinsky helped to update from the old treatises.

While some breeders keep pushing pigeons to extremes, Revolinsky follows the original standards as much as possible. Sometimes it's difficult, since fanciers were not so specific in the 1800s. For instance, the treatises say that the cere, a ring of skin around the bird's

F. Waller Lith.16.Hatton Garden.

ALMOND TUMBLER COCK.
1½ YEARS OLD.
2½ YEARS OLD.   4 YEARS OLD.

An illustration of Almond Tumblers from Robert Fulton's classic book shows how their dappled chestnut coloring changes with age.

The sleekly curved stance of Dan Revolinsky's Almond Tumbler would have driven Victorians wild.

eye, should be "fine." Does that mean a millimeter in width, or more? Drawings like Fulton's become his best guide—if they show a bird with a thinner cere or a smaller, rounder head, that's what he will aim for. "These are ancient birds," he said. "I'm trying to go backwards."

To see the fancy pigeons is to experience conflicting emotions. It's difficult not to absorb the excitement of fanciers and to get caught up in their aesthetic adventure. But it's also hard to look at a bird bred with its head thrown awkwardly against its back, or one that is so heavy it breaks its eggs, or one that can no longer preen its own feathers. It isn't *natural*. But Darwin pointed out that "monsters" are weeded out by selection, so these pigeons can't be thought of as monsters if their monstrosities help them survive. Success is dependent on context, and what would be considered aberrations in the natural world can lead to success in a world controlled by humans.

By the afternoon of the final day of the show, judging tables were being put away and the room was becoming oppressive after housing hundreds of pigeons for three days. I walked outside and sat on a bench outside the building. A pair of feral pigeons swooped across the bright winter sky, careened in a circle, and then settled atop the building opposite the show hall. Soon they were joined by five more, sitting in a row. They perched there for several minutes, absorbing the late afternoon sun that warmed the cold air. Then the pair took off again, beating their wings against an updraft so that they seemed to hover over the building for a few seconds before gaining speed and disappearing over the hall.

I felt palpable relief watching those pigeons fly free outside. Birds in flight have always been powerful symbols of transcendence. There's the rub: We want animals to be part of our world, to be under our control, and we also want to see them as untainted by humans. We make strong divisions between the animals that are part of our society and those that are separate.

While fancy pigeons took a high-speed course through evolution, these birds took a different route. Darwin was interested in the different fates of domesticated and wild animals. When he was studying pigeons, the birds fell into three major categories: wild rock pigeons that lived in remote areas, fancy pigeon breeds, and a large population of dovecote pigeons, some of which may have lived rather casually in cities. The more time they spent with humans, the more diverse pigeons became. The specimens of wild rock pigeons that Darwin was able to obtain seemed very similar to one another, varying only slightly among populations that lived in different places. Fancy pigeons, however, had completely transformed. Living in captivity made them dependent on humans; their diversification came at a cost. "On the view that all races are the product of variation," he wrote, "we can understand why they have not become feral, for the great amount of modification which they have undergone shows how long and how thoroughly they have been domesticated; and this would unfit them for a wild life."

Dovecote pigeons were a different story. Darwin found that they were more diverse in color and size than wild rock pigeons, but only slightly. There were no exaggerated anatomies to keep them from surviving in the wild. He had heard from naturalists in Madeira, Scotland,

and the United States of cases of dovecote pigeons going feral. Though Darwin himself did not bother to study dovecote pigeons, he might have shown more interest had he known what they would become. These birds hovered between wildness and domestication in a way the fancy birds didn't. They still had the ability to bridge two worlds.

5

Homing

In their history together, pigeons gained much from their association with people, and in turn people found ways to exploit pigeons for their own purposes. One of the first things people noticed about pigeons is their love of home, a quality humans have exploited to astounding ends. Even in ancient times, pigeons or doves were often seen as symbols of fidelity, sociability, and domestic life. It's a quality I began to notice even in the street pigeons in Boston, once I started to look.

Pigeons may be our very visible companions in cities, but we really see just a small glimpse of their lives, usually when they are bobbing about during the day looking for food. They are like people you pass downtown on your way to work, never knowing anything about their homes or families. Disconnected inhabitants of the city, both mundane and utterly mysterious. The life of pigeons, like ours, has two main arenas, work—finding food—and home. So rather than watching them forage in parks, a sight I was used to seeing, I began to look for their homes. I took walks around the brownstone apartment blocks and alleyways in Boston's South End where I lived, gazing up instead of watching the sidewalks. It wasn't exactly a voyage on the *Beagle*, but I did feel like an explorer, looking for the habitat hidden only by my own ignorance of it.

By the time I wandered into Blackstone Square Park, I felt as if I'd made a great discovery. A building facing the park, outfitted with columns and ledges, was home to a colony of nearly one hundred pigeons. Below a giant frieze reading "Municipal Building" were a row of columns, and near their capitals was a series of windows, crisscrossed with stone spokes. Behind the spokes were spaces where debris had piled up. These spaces and the ledge below them served as a series of "rooms" where the pigeons slept each night. It was a perfect place to watch their domestic life.

Pigeons are fond of routine. During the day, the birds largely abandoned their roosts, as they made occasional stops for a few minutes of rest on the upper ledges before taking off again on the hunt for food. But toward late afternoon, the pigeons slowly returned. First they would alight on the upper ledges or atop the decorated columns. Every once in a while, a noise or a flash of something in the sky would send the whole group of birds from the building in a shudder, to circle the park and then land again. As the sun waned and the park began to fill with dog-walkers and the rest of the after-work crowd, the pigeons began to make their way downward into the bedrooms. Always it was the same. They would fill the spaces along the windows, first the leftmost ones, and then, as the light faded, finally the rightmost spaces. They preened themselves and shuffled around for the best spot, gradually settling down for rest. Pigeons will not fly at night—their night vision is even worse than ours. Every night, they return to their chosen home to rest. They do it through the seasons. Some members of the pigeon and dove family do migrate, not long, dramatic migrations across continents, but more of a drifting to the southernmost extent of their range in the winter. But *Columba livia* is particularly inclined to stay put.

Some bird species are specialized for certain kinds of flight. An albatross, for instance, has enormous wings that allow it to glide for miles on air currents but keeps it from taking off and landing quickly. A quail can sputter straight into the air once it's startled, but it's usually exhausted its power after a journey of a few meters. Pigeons are generalists. They can glide a little. They can hover in the air for a second or two like a hummingbird. They can break into flight from a standing position, using a movement of the wings that scientists call clap-and-fling, which creates the distinctive fluttering noise when a flock of pigeons suddenly bursts into the air. If pigeons specialize in anything, it is maneuvering. A pigeon living on a cliff face can't hide in dense foliage—it is easily spotted by hawks and must be able to outmaneuver them as it flies to and from its nest.

Pigeons have something else in common with those people you see every day on your way to and from work. They are commuters. An albatross's wings are so large that flapping them for very long would exhaust it. Pigeons, on the other hand, are great flappers, and they can easily power their way through a commuter flight on a windy day. It's a lifestyle that people, once they settled into agrarian communities, no doubt related to—and took advantage of. Dovecotes would never have worked had the birds not been willing and able to return home each night. But people also found more extreme ways to make use of pigeons' natural behavior, to push the boundaries of their physical abilities and their deep attachment to their homes.

At dawn on a chilly spring morning in Rhode Island, Bob Rossi was getting his Homers ready for a training flight. I had met Rossi at a local pigeon show, where he was looking at birds to buy and race. He was a newcomer to the sport—it was only his third year flying—and at age thirty-three he was also the youngest adult member of his local club. He is at the point where he feels the need to prove himself to the old-timers but is not yet so caught up in the sport that he's obsessed with winning races. Instead, he is learning to purchase, breed, train, and manage his pigeons, mostly through trial and error.

Rossi's loft lay on an empty dirt lot at the end of a string of wooden houses. He greeted me, wearing a sweatshirt and baggy jeans, his shaved head tucked under his hood for warmth. The loft consisted of three small rooms he had constructed out of discarded pieces of lumber, half of it painted red and half just slabs of weathered fencing, with a small open area lined in chicken wire for the birds to fly around in.

Inside the loft, the pigeons began to coo. It started as a light murmur, but as Rossi entered one of the doorways and began to rummage around, it grew louder and rhythmic, rolling out in waves from inside the shed. Some pigeons began to fly out into the wire-bound area.

"This is like an aviary, they get to fly out here," Rossi explained, "but I can't let these birds out because they're prisoners."

They were birds he had bought from someone else, to use as breeders. Pigeons' attachment to their home loft is so strong that they will return to it even after living away for years. Rossi learned this lesson one day when he accidentally took one of the prisoners with his team of flyers for a training flight. The bird was the most expensive he had ever bought—it came with a certificate verifying it had won a race. When Rossi released the birds, he knew something was wrong: The whole flock went one way and a lone bird another. The pigeon flew all the way to its former home in Natick, Massachusetts, where Rossi was lucky enough to recover it later.

Perched against the wire was Rossi's first pigeon—a former street pigeon that he had taken in when she was injured. One evening when he stood outside a local bar, a young woman arrived, distraught, saying she had accidentally kicked a pigeon with a rock. Rossi offered to take care of it. The girl named it Rocky in honor of its injury, and its full name became Rocky Homer Pigeon Show. Rossi tried to release the bird once she was better; he set her in a tree, but hoped she would come back. She did. He credits the bird with changing his life.

Other pigeon breeders I talked to had similar stories about the moment they began to appreciate pigeons. Often, it began when they took a pigeon in as a pet and realized that even if the bird was set free, it would return to them. That the bird was returning for food or shelter didn't matter—it was coming back to *them*. Not only do pigeons choose to stay in their lofts, but with the proper training they will fly back even if their human caretakers repeatedly take them far away. And that's what pigeon racing is all about—it is the ultimate test of the bonds between people and domestic animals.

When the birds are young, the breeders begin training them first by letting them out when they're hungry; the birds learn to return to the loft for food. They begin to fly together as a team after repeated trips out. Eventually, the breeder takes the pigeons out in a crate—just a short distance at first—and lets them find their way home. The birds are taken on progressively longer trips, a mile, ten miles, twenty miles, and so on, until they are confident enough to return even from unfamiliar areas hundreds of miles away.

The first race of the season is next weekend. Rossi's birds, along with those of several other breeders in his area, will be loaded into crates and stacked onto a truck and taken out to Albany, New York, more than 150 miles away. The birds will be released together, and each bird, alone or with its loftmates, must find its way back home. Since the birds travel to different homes, the winner is determined by the time it takes to reach its home loft divided by the distance.

Rossi ducked into the room of the loft where his racers were nesting. Amid flapping in the fetid air, he began sorting through birds, pulling some into a wooden crate.

"If they're feeling sick, they'll feel weak; they'll let you catch them easily," he said. "They should be hard to catch, full of energy. That's one thing you gotta look out for."

He hesitated before grabbing one of the pigeons. He called her Big Momma, because she breeds so much. "If I take her today, she's pretty much going to be in the race next week. I think I'm going to take my chances. She's mated up with one of my really good prisoners, and they've got babies. I'm hoping the father will come and warm the babies up." He pulled one of the squabs out of its nest to show me—at just a week old, it was a scrawny, purple thing covered in bright yellow fuzz.

"At least there's two of them; they'll keep each other warm," he said. "If she doesn't make it home they'll probably end up dying, unless I take over the feeding." He paused. "But she'll make it home."

Learning the sport of pigeon racing takes trial and error—after all, these are living birds facing long distances, unpredictable weather, and the threat of predators they're not used to encountering. Rossi is constantly weighing his risks and trying not to make mistakes—in this case, weighing the need to train one of his strong fliers against her mothering duties. Once he had shoved about twenty birds into the crate, he carried it to a boxy black Lincoln Continental, which he had recently bought to provide the birds a smooth ride. He placed the pigeons in the trunk, where he had carved a hole into the back seat for ventilation. As we got into the car, he hesitated, then returned to the loft to check on Big Momma's babies. He came back to the car, muttering, "I just gotta hope that he comes down and hops on them babies. There's really nothing I could do—I have good feelings about her."

Very early in their history with people, pigeons began to make themselves useful as messengers. One of the first examples is in the biblical story of Noah's ark, in which the dove (or pigeon) returned to Noah bearing a branch that signaled it had found land. The previous messenger Noah had tried, a raven, never returned. Early Greek and Roman writings referred to doves serving as messengers of letters attached to their feet.

Despite the popular image of pigeons as scouts or messengers, from a pigeon's perspective it is only doing one thing: trying to get home. Pigeons are one-way communicators. But even a line of communication that runs one way can be useful, particularly when traveling away from a place with which you'd like to keep contact. In ancient Egypt, pigeons were carried upstream along the Nile to be released as a signal to those downriver that the floodwaters had arrived. In wartime, the same strategy could help convey messages from distant outposts to a central command. In many cases, pigeons could outpace messengers on foot or horseback, particularly over rough terrain. Julius Caesar may have used pigeons during his conquest of Gaul, and they also carried messages for Crusaders in the Middle Ages.

In later centuries, pigeons proved useful during sieges on cities. During the siege of Paris at the tail end of the Franco-Prussian War in 1870, pigeons were carried out of the city with refugees in hot-air balloons, where they were kept at lofts in London and other French cities. The birds were used to carry messages back into Paris on waxed paper rolls attached to their tail feathers. Later, messages were copied together, reduced in size by the new technique of microphotography onto small photographic films, then rolled into a goose quill and attached to the pigeons' feathers. Each pigeon could carry more than two thousand messages this way.

You would think that telecommunications would put pigeon messengers out of business, but the birds still proved useful in wartime— after all, wires get cut, and presumably it was more difficult to intercept a message borne by a flying pigeon than one delivered by telegraph or a tapped phone line. When World War I broke out in Europe, pigeons were well established as a means of reconnaissance in France, Belgium, and Germany. Later, Great Britain established its own pigeon service; the British air force would drop homing pigeons in baskets from planes as they flew over friendly territories, allowing those who found them to send information.

During the first and second world wars, pigeons were official members of militaries and honored as such. They became heroes in stories of bravery—in this version, the birds were not flying to get back to their nests and food but to save lives. In the U.S., the most famous wartime pigeon-hero was Cher Ami, a British-trained bird that was credited with saving nearly two hundred American lives in World War I. He was stationed with a battalion from New York that wandered into enemy territory at Grand Pre. Surrounded by enemy troops on all sides, they could not call for help with flares or by wire. So they sent their pigeons, all of which were shot down by enemy fire, save one. Cher Ami was struck by shrapnel that tore his leg, but he managed to make the forty-kilometer trip back to his home loft in twenty-five minutes. The message was credited with saving the entire "Lost Battalion," and the bird, now stuffed, perches on its single leg in a display in the Smithsonian in Washington, DC.

During World War II, Great Britain expanded its National Pigeon Service, in which birds were donated by breeders and were used to carry intelligence from Europe. The United States also expanded its pigeon operation: At its peak, the U.S. Pigeon Corps had three thousand enlisted men, one hundred and fifty officers, and fifty-four thousand pigeons. They were divided into companies—they had names and numbers, and their wartime exploits were recorded. In 1943, Britain established a medal for bravery in animals, many of which went to pigeons with names like Flying Dutchman, Beachcomber, Commando, William of Orange, Billy, and Princess. One of these recipients was G. I. Joe, a famous pigeon from World War II, who hatched in North Africa and was trained by American forces and stationed at a British headquarters in Italy. In October 1943, an infantry division requested air support in taking Colvi Vecchia, a town held by the Germans. But when the troops suddenly entered and seized the town, G. I. Joe was sent to stop the bombers, arriving at the headquarters with his message just minutes before the planes took off. Militaries had some success using portable lofts that pigeons could be trained to return to even when moved. There was even a system for training "two-way" pigeons that could fly to two locations, one that the bird recognized as the home loft where it slept, and another where it could get food.

Networks of these pigeon-messengers were also joined together to create a pigeon post. Pigeons were participating in large-scale communications networks as early as the fifth century BC, when a pigeon post was established in Persia and Assyria under the reign of Cyrus the Great. In the twelfth century, a large-scale pigeon post ran from Baghdad to Syria and Egypt, and two hundred years later, Turkey carried a network of pigeon towers spaced at forty-mile intervals. In nineteenth century Europe, pigeons found a new niche: delivering highly valuable stock exchange information in the days before the telegraph was widely used. A few speculators in London began running pigeons between London and Paris, giving them access to information well ahead of the general news. These "pigeon men" were known to have an advantage, and their appearance at the market was sure to forecast a rise or fall. Eventually, some of the financial brokers refused to do business until foreign newspapers arrived to report the news to all. In Belgium, Paul Julius Reuters, the founder of the famous wire service, first established his reputation by flying pigeons bearing news and stock prices between Brussels and Aachen, Germany, where there was a gap in telegraph communications, besting the railroad's time by two hours.

A British army detachment releases a two-winged messenger during World War II.

Having money at stake increased the pressure to find pigeons that could deliver messages quickly and accurately, particularly because breeders could rent out their birds for work. In the early 1800s, Belgian fanciers experimented with different crosses of pigeon breeds to find a better flyer; they created a breed called the Anversois and another called the Smerle. By 1850 these, along with crosses from several English breeds, were fused into one variety, the Belgian Racing Homer. The English also began to breed special homing pigeons after pigeons made a splash in the siege of Paris. Those exploits helped to popularize the idea of racing pigeons, and the new sport quickly spread. The new breed of Homers were bigger and more muscular, allowing them to accomplish longer flights home, up to several hundred miles in one go.

Rossi and I drove west along Route 6. The orange light of dawn sheared the road as we traveled past wooden houses, thin woods, a lake, and, after several minutes, a soccer field that Rossi uses as his ten-mile release point. Finding a good site is important—he looks for large fields, preferably on high ground and far enough from the roads that the birds won't careen into traffic as they pour from the cage.

Competition is high in pigeon racing—there is prize money at stake in races, and usually opportunities for betting on each bird. And of course, there is the simple fact of winning. It is not the gentleman's sport that pigeon fancying is. The racers are reluctant to divulge secrets about how they train and care for their birds; some have elaborate routines for preparing their birds for races, and give the birds special foods, medicines, and herbs. It's difficult for a newcomer like Rossi to get advice, and it's even harder to get good birds. Some breeders will talk up a bird's performance to get a sale, while the ones with really good stock, he said, "would rather kill their birds than sell them to a guy like me."

The woods thickened, and we turned onto a two-lane road that Rossi said was the route most of the racers in the area followed, though few of them admitted to it. "On the map it's a straight line to Albany," he explained.

A spring Nor'easter was scheduled to arrive that evening, but for now the sky was clear. Rossi has learned to pay attention to the weather. The previous year he lost most of his birds when he flew them in a race and a storm hit. In a race, birds often fail to come home, but every once in a while there is a "smash" in which nearly every bird vanishes. After that smash, he began attending to the weather forecast, and avoided flying the birds anytime a storm threatened. "That one bad day, if you can avoid that it's so big," he said ruefully. "Every day you can avoid losing that bird is huge because—you can only lose them once, really."

Aside from his work in construction, Rossi's life revolves around his pigeons. He visits the loft, a few blocks from his house, up to four times a day during this time of year. In the weeks before racing season begins he is out training them on nice days, usually starting before dawn. The hobby has kept him from staying out late drinking, and has let him see more of the countryside—at least the Albany-to-Providence line, of which he now claims to be a master.

We passed the twenty-mile release point and the thirty-mile one. Rossi had driven nearly forty miles when he saw a grassy field to the left of the road that looked like a good place to stop.

"It's 7:07, that's a good sign," he said, indicating the clock on the dash.

He parked in a dirt pull-off and hefted the crate out of the trunk. It was sunny now, with a light breeze. A few geese stood in the empty field; surrounding it were tree-covered hills. Rossi carried the crate midway across the field and set it down. Through the wood slates at the top of the crate, I could see the birds shifting against each other in agitation.

He lit a cigarette, saying he wanted to give the birds a moment to collect themselves. But it was probably as much for him as for the pigeons. Pigeon racing, despite its rewards, also gives him a lot to worry about. Figuring out which birds to buy, how to pair them, how to doctor their illnesses, which birds to fly and which to breed, how best to train, when to let them rest—it's all about weighing risks. Every release is a gamble, whether there is money on the line, or glory, or just the birds themselves. There might be hawks or winds, or just the chance that the pigeons will lose their way. Every release involves letting go.

After a few minutes, he bent down and placed his hand on the latch at the side of the crate. "You guys ready?" he said. "Good luck!"

The door opened. The pigeons rushed the opening, brushing against the sides of the crate and spilling out in a flurry of feathers. They swooped, en masse, low to one side and then the other. They began to execute figure eights in the air, growing higher with each one, moving away to one end of the field, then drifting toward us again.

"They're coming back; they're taking a good look around," Rossi said as the birds drew a wide arc above us. After this last time circling, they suddenly shifted to the southeast, the direction of the loft, and flew straight away.

"Now they're pretty much on their way home."

Their bodies became small chevrons and then just undulating specks. Then they disappeared entirely, as if swallowed by the morning sky.

Though pigeon homing has been exploited throughout history, it wasn't until the nineteenth century that people began to study the question of how pigeons manage to find their way home from faraway and unfamiliar places. Two main theories emerged. One was that pigeons remembered the route of their outward journey and simply retraced it, based perhaps on landmarks. Another theory was that pigeons have an innate sense of their home, based on some sort of inner compass. Both of these turned out to have some truth to them.

At the beginning of the twentieth century, scientists began to study the homing abilities of pigeons released at different distances from their lofts. At the same time, research uncovered that wild seabirds like petrels and shearwaters could find their way home across open ocean without any need for landmarks, suggesting that all birds have an innate sense of direction.

In the middle of the century, a couple of developments helped to boost the science of bird navigation. One was methodological: Researchers found that the direction in which birds disappeared from view, or their vanishing bearings, roughly corresponded to the route they took to get home. It may seem like an obvious point, but the ability to simply calculate vanishing bearings made it much easier to get around the huge problem of tracking birds in flight and to collect data about their behavior. Another advance was theoretical. A German scientist named Gustav Kramer pointed out that homing involves solving two problems. One was the problem of sensing direction, an ability he called the compass. But knowing north from south is useless if you don't also know position—where you are in relation to home. He called this ability the map. The map-and-compass model provided a framework for thinking about different aspect of navigation. Bird navigation became an intensive area of research, and homing pigeons became its primary subject; it was thought that studying these birds, even though they were bred and trained by people, could shed light on how wild birds migrate.

The best way for scientists to determine how pigeons find their way home was to attempt to thwart them. And so, over the next decades, homing pigeons raised in university lofts were subjected to all sorts of strange challenges. They were released wearing frosted goggles that made it difficult to see landmarks. They were raised in enclosed spaces and exposed to unusual light cycles to "clock-shift" them into believing it was a different time. They had magnets attached to their backs, and battery-powered magnetic coils affixed to their heads. They had nerves cut to remove their sense of smell. They were anesthetized on the journey to their release sites and given only recycled air to breathe.

All of these studies—and there have been hundreds—have provided insight but also created mysteries. Figuring out the compass was the clearer problem. Pigeons have at least two ways of establishing direction: by looking at the position of the sun in the sky and by sensing the earth's magnetic fields. The magnetic sensing was recently found to be the product of iron-containing particles inside cells of the birds' upper beaks.

Studies of the map problem have been trickier, and every few months another paper seems to add a new wrinkle. Scientists have split in different camps, some believing that pigeons rely on their sense of smell to navigate, others who think the birds just follow roads and landmarks. It could be that all of them are right. That's the conclusion Charles Walcott, a biologist at Cornell University, reached. Walcott was in the thick of navigation research until the mid-1990s, when his pigeon loft at Cornell was shut down by administrators fearful that pigeons were unsanitary. Before then, Walcott conducted all sorts of experiments with pigeons in New York and Massachusetts to see what influenced their homing abilities. The problem, as he put it, is that no matter what scientists do to pigeons, "the little stinkers all come home."

Some come home even if they have been clock-shifted and then released under cloudy skies so that they can't use the sun. Others come home if their magnetic sensors are thwarted with magnets. Whenever a research group publishes a paper showing they have interfered with pigeons' ability to home, another group's data contradict it. For instance, Italian pigeons allowed to breathe only filtered air on the way to their release site did much worse at finding their way back than pigeons that could breathe fresh air, leading the scientists studying them to conclude that smell was the most important factor in homing. But Walcott's pigeons in Ithaca had no such problems. "In this part of the world, if I take a homing pigeon and I anesthetize it, I put it in a sealed container fed with filtered bottled air, on a rotating turntable, and I drive it out a hundred miles or so—when it gets over being carsick, it flies straight home."

Walcott believes that the answer may be something that scientists don't usually like to admit: individual variation. For convenience's sake, scientists like to assume that their subjects are interchangeable, that they have universal ways of doing things. But Walcott now believes that pigeons have a set of navigation tools that they can draw upon, but which tools they rely on depends very much on their individual upbringing. The pigeons in Ithaca and Pisa may react differently because they have different experiences.

That's the only way to explain one particularly frustrating phenomenon that Walcott encountered. When he was studying pigeons at a loft at Fox Ridge Farm in Lincoln, Massachusetts, Walcott and his colleagues found that pigeons released at a magnetic anomaly under sunny skies were disoriented—the bigger the anomaly, the more disoriented they were. The findings were exciting, so Walcott's team wrote up the results in a paper and had it published. But when he moved to Ithaca and tried the same experiment with pigeons raised at Cornell, he experienced what every scientists dreads: a contradiction. The Cornell birds actually oriented themselves a little better at magnetic anomalies, which seemed to disprove his previous work.

Walcott decided to reestablish the loft in Lincoln, but because the old loft was no longer available, he found another site at Codman Farm a mile away. He stocked it with pigeons from Cornell and Boston; none of them seemed to care about magnetic anomalies. He then persuaded the owner of Fox Ridge Farm in Lincoln to allow him to set up a portable loft on the original site. As before, he found that pigeons raised at Fox Ridge Farm were totally disoriented by magnetic anomalies, though the birds raised a mile away were not bothered in the least. "It's clear that it isn't the pigeon stock, it's the place where they were being raised that mattered," he said.

Magnetic fields coming from the ground are not totally uniform—

they vary depending on local features, like the presence of metal in the earth. When his team looked more closely at the magnetic makeup of the two sites, they found that Fox Ridge was located in a place where the magnetic field changes dramatically, while at Codman the field was relatively stable. If you were to map out the regions magnetically, the Fox Ridge loft would be perched on the side of a magnetic mountain, while the loft in Codman Farm would be nestled in magnetic flatland. Walcott's hunch is that the pigeons of Fox Ridge Farm learned that magnetic fields provided useful information about where they were, just like a person growing up on a slope might use uphill and downhill to tell direction, while the pigeons living on Codman farm would have learned to rely on other cues.

In his view, as pigeons grow up and begin to fly and sense their surroundings, they acquire a lay of the land. They learn the physical landmarks, including roads. They smell the odors in the air. They sense the magnetic field and learn the path of the sun. All of it combines to form a mental map of their environment, which they use when they are finding their way back.

After Rossi's birds disappeared from view, we got back into the Lincoln and retraced our path along the thin highways back to his loft. Rossi believes his birds do follow roads, but because roads can be obscured by trees in this heavily wooded region, there's a theory among the local racers that pigeons follow power lines instead. On a couple of occasions, Rossi has spied his flock circling over the field at the ten-mile release point while he waited at a red light. But for the most part, what happens between release and return is a mystery.

Nowadays, GPS technology has allowed scientists to start filling this gap. The research supports the racers' notion that birds follow cues on the ground. While pigeons might use compasses to orient themselves in unfamiliar places, when they are flying in familiar territory they rely on visual landmarks. They might fly from one noticeable feature to another—a behavior called "steeple-chasing"—or they might follow a river, road, or railway track in roughly the direction of home. The term "as the crow flies" doesn't seem to be accurate, or at least not for pigeons. Each individual bird develops a preferred route home, but it rarely follows a straight line.

GPS has also allowed scientists to begin studying the social context of flight—whether flying in the wild or in races, pigeons usually fly in pairs or flocks, so choosing a route is rarely an individual decision. A team at Oxford found that certain birds tend to determine the course of the group, while others are passive, willing to deviate from their course if it means staying with the others. When pigeons from two different lofts were released together in pairs, scientists found that most of the pairs stayed together for much of the journey, often following the route favored by one of the birds, only breaking up at the end to make the final stretch alone.

In the end, what matters is not just the ability to find their home, but the desire to return. Racers know this and do all sorts of things to keep their birds motivated to return home, from starving them to placing them in a pen with a potential mate right before the race and then taking them away before the courtship can be consummated, a technique called "widowing."

And that sort of motivation might be the key difference between the average street pigeon and a trained homing pigeon, beyond the latter's pumped-up physique. People who breed homing pigeons often disassociate their birds from ordinary street pigeons—a pamphlet from the American Racing Pigeon Union reads: "To place the modern Racing Homer in the same category as the common street pigeon would be analogous to placing the thoroughbred racehorse in the same category with the plough nag!" But given that the Homer line goes back just a couple of hundred years, a blip in pigeon evolution, it's doubtful that they have navigational abilities that other pigeons don't. Both feral pigeons and wild rock pigeons have the ability to find their way home, though studies have found the ferals are not as successful as Homers. Part of the reason is that they haven't been trained to be faithful to their lofts; they are more likely to stop to eat or join up with other flocks along the way.

The paradox of homing pigeons is this: Their owners exploit a natural tendency, their faithfulness to home, to get pigeons to behave as travelers. In some sense, every race is a forced migration. But whether the birds learn to enjoy their journeys is impossible to know.

Today, we were lucky with weather; with a light wind at their tails, the pigeons easily beat us. When we pulled into the dirt lot in front of the loft, a few birds lingered on the roof, and two circled overhead. Rossi identified one of them as his most consistent flyer. The bird's flight was more confident and purposeful than a feral pigeon. It looked like a professional runner taking an easy lap around the track after a race, as if it were enjoying a last moment of exercise before landing. The others, Big Momma included, had already passed through the swinging trap doors at the top of the loft, back to the safety of home.

6

Hunt and Peck

At the same time that pigeons were serving as messengers in World War II, a covert project was employing the birds in a different wartime service, bearing missiles instead of messages. On a train ride from Minneapolis to Chicago in 1940, psychologist B. F. Skinner was watching a flock of birds maneuvering in formation in the distance. With the war on his mind, he began to see them as small navigation machines, and he wondered if it might be possible to harness the skills of birds in solving the problem of guiding missiles to targets. European cities were being devastated by aerial bombing, against which they had little defense, and there was no way to target missiles against the attacking planes.

Skinner would eventually become modern psychology's most controversial figure, but at the time he was still a relatively junior researcher in the field of behaviorism. That burgeoning offshoot of psychology emphasized studying the outward behaviors of animals and people and the way their environment affected those behaviors. In doing so, it eschewed explanations related to inner thoughts or personality in favor of observable causes in the outside world. In his graduate and postgraduate work at Harvard University, Skinner had developed a simple but powerful apparatus for studying behavior in rats. The animal was placed in a box with a lever that it could press to receive food. The food acted as what Skinner called a reinforcer of behavior: Whenever the rat received food after performing some action, it was more likely to repeat that action. By changing the way the rat was rewarded with food, Skinner could "shape up" complex behaviors—in one instance, he taught a rat to do acrobatic tricks simply by feeding it at the right moments. Previous work with rats focused on mazes, and the behaviors of the animals were varied and unpredictable. By simplifying the environment to a simple box with just the feedback of lever-pressing and food reward, Skinner could get very predictable results from his rats. It opened up the possibility that behavior was not random or eclectic, even in complex situations in higher organisms, including humans.

Skinner was single-minded and often brash in his assertions. *The Behavior of Organisms*, published in 1938, presented his view that behaviorism was the future of psychology, and contained very little acknowledgement of past theories and other work. And although he had to that point spent his time shut away in laboratories, the war was pushing Skinner to look for ways to apply his discoveries in the real world. And so, when Skinner looked at those wheeling birds, he didn't see free and inscrutable wild creatures but navigation machines that, with the right reinforcement, could be recruited for a higher purpose— the end of the war. And thus Project Pigeon was born.

Skinner bought a few pigeons from a local poultry shop and experimented with turning the birds into pilots. He found that if he slipped a toeless sock over the pigeon's body to restrain the wings and feet, the pigeon could use its beak to peck a target and pick up grain delivered as a reward. The movements of the pigeon's head and neck could also give steering signals toward the target. Working with his students at the University of Minnesota, Skinner built a crude model of a pigeon-guided steering system. Since he was a child, Skinner had always tinkered and built contraptions of his own design. He and the students built a hoist to which pigeons, restrained in "snuggies" made of socks

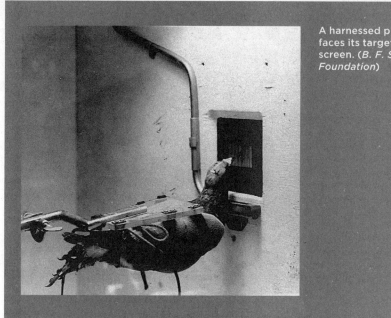

A harnessed pigeon faces its target on a screen. (*B. F. Skinner Foundation*)

and pipe cleaners, were tied; when the birds pecked at a bull's-eye in front of them, their head movements aligned the hoist with the target with electric motors. They demonstrated the system to the dean of the school, noted physicist John Tate, who passed a description of the project to the National Defense Research Committee. The strange proposal was rejected.

Skinner returned to the idea again the following year, after the attack on Pearl Harbor. His initial idea was to use pigeons to guide defensive missiles that could be dropped from high-flying planes onto bombers below. Now, he switched to an offensive missile, which would hold a sacrificial pigeon that could guide the bomb directly to a target on land. Skinner trained pigeons to steer a contraption toward a target on a revolving turntable. He made a film showing the pigeons hitting small model ships, and again sought funding. Despite some polite encouragement, Skinner found no help in government. But he finally found support in an unlikely place, the General Mills Company. Arthur Hyde, the company's vice president of research, took the project on as a public service rather than a potential product. Skinner and two students moved to the top floor of an old flour mill with a $5,000 grant to pursue the project in earnest.

It seems unbelievable that pigeons, which were not even considered particularly smart birds, could be entrusted to guide bombs. But in fact they were highly reliable pilots. Using his principles for shaping behavior, Skinner could quickly teach the birds to peck a target with great precision. Pigeons have excellent visual acuity, and the team trained the birds to focus on targets like images of ships or buildings or a particular street corner, all while ignoring other distracting targets or clouds.

Skinner found that if they were kept just a bit underfed, the birds would work tirelessly for their reward. They could perform under changing conditions, resolutely following their task with little regard for noise or pressure changes akin to what would be felt in a dropping bomb. It seemed likely that a pigeon, properly trained, would keep pecking its target even as it hurtled toward certain death from miles above the ground—a situation that would be eerily mimicked by the Japanese kamikaze pilots who would soon decimate Allied targets, except that the birds were only after a meal, not glory. It wasn't even necessary to feed a pigeon that often. A bird would peck for several minutes if it thought that food would come in the end—one of Skinner's birds made ten thousand pecks in forty-five minutes without stopping. In the early stages of assembling the device, Skinner's team was far more concerned with the mechanical challenges—how to translate the pigeon's signals into movements in the missile—than it was with the reliability of their pilots.

In fact, the primary behavioral challenge for Project Pigeon was not in guiding the actions of pigeons but rather the human sponsors and evaluators. With the work at General Mills, Skinner produced another demonstration film and won a $25,000 federal contract to develop an "organic homing device." But his team had difficulty getting information from the government about the missile the pigeons were to steer, which was under separate development. The pigeon pilots would sit in a cone-shaped missile called the Pelican, which was a glider that had not yet been steered successfully by any other homing device. It was being tested on a strip of land along the New Jersey coast, and Skinner was given color photographs of the site taken from various angles, which he used to train the pigeons. But he never received precise information about how the steering signals from the pigeons would interface with the Pelican, so his team had to make arbitrary decisions about how to refine the signals as they designed their homing device.

Skinner expanded the missile to include three pigeons, therefore building some redundancy into the system with a new psychological pressure: If one pigeon got off track and started pecking a different target, it would stop receiving food, thus pressuring the birds to reach a consensus and follow the majority opinion. A few months into the new contract, Skinner demonstrated the device to the defense committee. Unfortunately, the guesses about the design of the missile that the researchers made turned out to be the wrong ones, the data they produced did not look as impressive as they thought it could.

Skinner believed that all these problems were technicalities that could be fixed with time and the right information. A few days later, his group was given one last chance to make their case in front of a committee. "By this time we had begun to realize that a pigeon was more easily controlled than a physical scientist on a committee," he later wrote. "It was very difficult to convince the latter that the former was an orderly system."

In a country enthralled with technology—the same country that would soon unveil an atomic bomb—the idea of using pigeons as warfare agents must have seemed as atavistic as sending troops into battle on horses instead of in tanks. The best thing Skinner could do was focus on the reliability of the results.

But again he was thwarted. In order to convince the wavering committee of the infallibility of his project, he set up a demonstration. The researchers placed a pigeon in a black box with a translucent window on which a projector cast an image of its familiar target in New Jersey. As the pigeon pecked, Skinner planned for each scientist in turn to observe the behavior by peering down a tube into the black box. But pressed for time, Skinner was asked to take the top off the box so that everyone could see at once. Even though the bird had been harnessed for thirty-five hours and might be distracted by all these onlookers, the pilot pecked with machine-like precision. But the effect was lost. The curtain had been lifted, and the wizard behind Skinner's navigational machine was just a hungry pigeon.

"It was a perfect performance, but it had just the wrong effect," Skinner remarked. "One can talk about phase lag in pursuit behavior and discuss mathematical predictions of hunting without reflecting too closely upon what is inside the black box. But the spectacle of a living pigeon carrying out its assignment, no matter how beautifully, simply reminded the committee of how utterly fantastic our proposal was. I will not say that the meeting was marked by unrestrained merriment, for the merriment was restrained. But it was there, and it was obvious that our case was lost."

With this meeting, pigeons reached a limit in their usefulness to humans. Even though they had won medals for serving as messengers in the war, they would never be entrusted with guiding missiles. Perhaps the difference was in the action; we don't mind letting animals perform tasks that humans can't, like birds flying with messages or dogs sniffing out drugs in the airport. But try putting a bird in a cockpit or a dog behind an X-ray scanner, and suddenly a noble task is just a situation for ridicule. And a pecking pigeon in particular lacks gravitas. As Skinner left the meeting, Arthur Hyde offered this advice: "Why don't you go out and get drunk!" Skinner soon received word that further funding was rejected. After two years of intense work, all he was left with was "a loftful of curiously useless equipment and a few dozen pigeons with a strange interest in a feature of the New Jersey coast."

But Skinner had gained something else. He had begun to see how behavioral research in the lab could work in the real world. Behavior was typically seen as something ruled by whim, fancy, fate, or the mysteries of the soul, but Skinner had shown that, on the contrary, behavior could be predictable and controllable. Psychological principles could guide a missile to its target as surely as the laws of physics. Skinner had found the animal that he would rely on for the rest of his career. He stopped working with rats and focused his studies on pigeons.

In addition to working with pigeons while developing his guidance system, Skinner had also come into contact with the birds while working in the General Mills flour mill, where large flocks of wild pigeons

Figure 1. Thirty-two
Trained Birds in
Jackets.

Figure 2. Front of Apparatus, Showing Boxes Holding
the Harnessed Birds.

Figure 3. Rear View Showing
Camera Obscura Projectors

Figure 4. Close-up of Single
Panel. Fixed Ratio at Right,
Reinforcer at Left, Photo-
Relay at Lower Left.

Photos from the Project Pigeon files show the jacketed pilots and their training modules.
(*Harvard University Archives, call # HUGFP 60.50, Box 2*)

gathered to feed off spilled grain. All day, pigeons circled and crowded around the windows, and the scientists would sometimes snag a bird from the sill and begin to tinker with its behavior the way they would an experimental apparatus. First they built a food magazine that would make a noise and then automatically dispense grain into a cup. The device helped them train the pigeons because the sound of the magazine gave an immediate signal to the birds to let them know that food was on the way, like the "correct answer" bell of a quiz show. The scientists used the device to shape behaviors in the birds, like getting them to peck so hard their beaks would hurt.

One day, they decided to take on the more ambitious task of teaching a pigeon to bowl. In a box, they set up a wooden ball and a set of toy pins. They placed a pigeon in the makeshift alley and waited to see if it would figure out how to roll the ball with a swipe of its beak.

The pigeon did nothing. Skinner's principles of reinforcement depended on a subject making a behavior that could be reinforced—it could only learn by first doing. Impatient, the scientists began to feed the pigeon whenever it did something remotely similar to the performance they were after—even if it just looked at the ball or made some swiping movement with its beak. The result was astounding. "In a few minutes, the ball was caroming off the walls of the box as if the pigeon had been a champion squash player," Skinner wrote. A pigeon would rapidly adapt its behavior in response to food, accomplishing tasks that no pigeon would seem capable of learning in the wild. The experiment perfectly illustrated that even complex behaviors could be shaped through incremental reinforcement.

When General Mills ended Project Pigeon, Skinner built a dovecote in his garden and kept the birds around for experiments of his own fancy. But he was eager to get back to serious study, and when he accepted a position at Indiana University he reconfigured his lab equipment for pigeons.

Most of our current knowledge of biology comes from a select group of species—laboratory animals, or model organisms. These species, which include rats, mice, worms, fruit flies, certain species of fish, frogs, and rhesus monkeys, are the workhorses of science. While exotic wild species are the subjects of nature documentaries and exciting research in ecology, the model organisms are the source of the most basic findings about how life works. Domestic pigeons have often been a convenient subject for laboratory scientists. The hormone prolactin, for instance, was discovered in pigeons; it stimulates milk production in their crops the same way it stimulates breast milk in mammals. Much of what we know about bird biology in general comes from studies using pigeons, including how they metabolize food and how their sensory systems work. Pigeons flown in wind tunnels and with sensors attached to their large pectoral muscles have revealed information about the aerodynamics of wings and the energy costs of flight. Homing pigeons, of course, have helped people study navigation, and a few scientists since Darwin have also taken an interest in fancy pigeons for their color genetics. But in psychology research, lab rats and other mammals were very much the species of choice before Skinner.

Skinner found that pigeons were well suited to laboratory life. Having been domesticated, they were used to living in cages and fraternizing with humans, and they were more resistant to diseases inherent to living in large groups. As birds go, they were a good size; they didn't have the fluttering metabolisms of small songbirds that barely cling to life, nor were they large and cumbersome like geese. They were abundantly available and cheap to procure and feed. Skinner also found that they were well suited to studies of reinforcement. He redesigned his boxes and gave the pigeons round keys to peck rather than levers to press, since Project Pigeon had shown that the birds were far more dexterous with their beaks than their feet.

Skinner saw the shaping of behavior as a kind of evolutionary process. Reinforcement from the environment shaped an individual's behavior just as natural selection shaped the genetic makeup of a species. Even the nuanced behaviors of people could be explained by their environment—the sum of rewards and punishments over a lifetime. Skinner never argued that human behavior was as simple as that of pigeons and rats. But scientists had often found that basic principles of physiology are similar from "lower" organisms to "higher" ones, the differences being of complexity rather than kind. The same could be true for behavior.

Skinner's studies showed how organisms were always adapting

their actions to their environment—even to the point of drawing false connections between the two, which Skinner called superstitious behavior. If a pigeon was fed every fifteen seconds regardless of how it behaved, the bird would still connect the feeding to whatever it was doing just before the food came. One pigeon began turning around several times between each feeding, another pushed its head against a corner of the cage repeatedly, and two others swayed from side to side. It was easy for Skinner to draw a parallel to the baseball player's rituals at bat or people's persistent beliefs in good luck charms and even religious ceremonies.

For the most part, though, Skinner wasn't interested in the behaviors of the birds per se. Instead, he was interested in the relationship between behavior and reinforcement, because he believed that all behaviors would follow broad principles. At Indiana and later at Harvard, Skinner focused on the rate of response—how often a pigeon pecks a key in response to different kinds of reinforcement. Would it peck more often if it received food after each peck, or every ten pecks? How long would it take a pigeon to give up pecking once the food supply stopped? He and the students in his lab measured the number of pecks and intervals between pecks, all of which could be easily plotted out. Eventually, much of the work was published in a heavy, graph-laden book called *Schedules of Reinforcement*.

Though the data might seem dry and pointless, to Skinner and the young scientists who came to the Harvard Pigeon Lab, they were incredibly exciting. The idea that behavior could be quantified and plotted in neat, predictable curves was revolutionary. The lab, led by Skinner and later Richard Hernstein, was alive with activity. Students spent most of their waking hours in the lab and gathered in weekly staff meetings where research projects were loudly debated. They were encouraged to tinker, experiment, and improvise, only loosely supervised. Their excitement was not from uncovering the inner secrets of the mind, but discovering that something *works*. To be able to devise a hypothesis and see it confirmed in the behavior of a living animal was exhilarating.

To the rest of the world, though, behaviorism seemed too narrow. Skinner in particular was a divisive figure—he took his ideas to extremes and dismissed concepts such as free will that were fundamental to American culture. His view that behavior was determined seemed threatening and even ridiculous. In scientific circles, few were willing to treat the mind as an unknowable black box as Skinner did. By narrowing his interests to observable behavior, Skinner had achieved exactness, but for many scientists the price was too high. *Schedules of Reinforcement* contained graph after graph showing what happened when a pigeon was rewarded in one way or another, page after page. But it had little to spark the imagination, to suggest a meaning behind the data, and the book is rarely cited today. For most psychologists, neat data sets were not worth abandoning the quest to understand the workings of the mind.

Eventually the Harvard Pigeon Lab folded under the growing "cognitive revolution" in psychology. The emerging science of computers and artificial intelligence made it possible to explore and test ideas about inner processes and states in the brain. Strictly behavioral research continued at small scattered labs, and behaviorists have continued to hold meetings and publish journals. They formed the Society for the Quantitative Analysis of Behavior, its acronym, SQAB, an homage to the bird that helped launch the field.

Skinner had always looked for ways to apply his ideas in human society, even envisioning utopian societies in which behavioral engineering solved social ills. Behavioral reinforcement did work in society, though not in the all-encompassing ways Skinner envisioned. Few people today call themselves behaviorists, but its approaches filtered into business, education, and science, and it has drawn renewed interest from clinicians looking for ways to manage autism. Its impact was felt in the lab as well. For psychologists, even cognitive ones, it was no longer acceptable simply to contemplate the mind—they now had to measure behavior and collect data to make a point. Behaviorism made psychology more rigorous. In all of this change, laboratory pigeons were not abandoned. Their history in psychology research assured them a place in labs of cognitive scientists.

As for Skinner's earlier dreams of turning pigeons into functional machines, on the few occasions when people tried to harness the power of pecking for useful purposes they met a similar fate. Thom

Verhave, a psychologist working at a pharmaceutical company in the 1950s, tried to apply pigeons to an onerous aspect of commercial production: quality-control inspection. Touring the area where capsules were manufactured, he watched as about seventy women examined each capsule one by one on conveyer belts, discarding the "skags," capsules that were dented, misshapen, or discolored. Figuring that pigeons could do the same task, Verhave proposed the idea to the director of research, who had just managed an expensive but failed attempt to use a machine to inspect capsules. Verhave was given permission to develop a demonstration device, and he trained two pigeons to recognize common defects and report if they spotted one. Within one week of training, both birds were inspecting capsules with 99 percent accuracy. Their performance garnered visits from higher-ups at the company, but soon the board of directors quashed the project. Even if it worked, they argued, who would buy drugs from a company that used pigeons for quality control?

Another scientist contracted with a company to train pigeons to inspect diodes on an assembly line. But once it was clear the project could be successful, the company quickly dropped it. Another scrapped project was an attempt, in the 1970s and 1980s, to train pigeons to do search and rescue from Coast Guard helicopters. The birds could identify floating objects in the ocean with a much higher success rate than humans, even with hours between sightings, but the project never came to fruition. Time and again, humans overseeing the projects turned out to be as predictable as the pigeons.

On the wall in Robert Cook's office at Tufts University hangs a pigeon skull, mounted and framed, a tiny thing. The back casing once held a brain smaller than a thumb tip. The front is dominated by a gaping whorl, rimmed in eggshell-thin bone, where the eye sat.

"Their head is mostly eye," Cook remarked. Their eyes take up so much space inside the head, in fact, that just a delicate membrane of bone separates them. If we had comparable eyes, they would be the size of softballs.

A large part of a pigeon's brain is devoted to vision. This makes pigeons particularly good subjects for Cook's research, which focuses on visual cognition, or the way the brain processes what it sees. The visual information that flows into the brain from the eye's retina is little more than pixels of varying color and shading, like the information from a digital camera. The brain must somehow interpret that raw information, turn the pixels into objects that have depth, shadow, boundaries, and movement.

Cook, who has an amiable, teddy-bearish demeanor, is one of the cognitive scientists who took up the pigeon as a subject after behaviorism fell from fashion; he calls his field animal cognition, a term that Skinner would have considered an oxymoron. Today, scientists are willing to accept that animals have complex brains that do many of the same things we do, even forming concepts and solving problems. Rather than dismissing the brain as a black box, cognitive scientists like Cook try to figure out its inner workings through a combination of theory and experiment. "In some sense, we're trying to reverse engineer the organism, to find out what its capacities are," he said. Once they know the pigeon's capabilities, they can start to hypothesize how the brain processes information and test those hypotheses.

Bird brains pose an interesting case study, since they must be particularly streamlined. Brains are expensive—they are heavy and guzzle energy—and a bird, whose entire body is adapted to be as light as possible for flight, doesn't want more brain than it absolutely needs. In songbirds, the brain region devoted to song even grows and shrinks seasonally as it is needed.

Sight must be critical to pigeons to warrant such brain space and weight. Pigeons undoubtedly see the world differently than we do. Though they lack our depth perception, pigeons have better color vision; while we have three different types of color pigments in our retinas, they have five, including one that detects ultraviolet light. Our eyes have one region of focus where we can refine small details, at the center of our vision, but pigeons have two areas of focus, one in front and one out at the sides of their vision—probably useful for detecting predators in the air even while eating on the ground.

Pigeons are not considered "smart" birds; some biologists I talked to dismissed them as notoriously stupid. They don't play games or solve puzzles like ravens and crows do, traits we usually associate with intelligence. But scientists have uncovered lots of surprising abilities, particularly in how they see and remember. "One of the things that has motivated what we do is to understand how a small-brained creature can do so many different things," Cook said. "Vision in particular, but we've also found that they have pretty prodigious memories and can even form some concepts. It's rather remarkable what they can do."

Cook's pigeons, about fifty of them, live in rows of stacked metal cages, each marked with a name bestowed by one of his graduate students: Oprah, Geraldo, Montel, Nemo, Bilbo, Brady, and so on. They are large, stately birds, not street pigeons but Silver Kings and White Carneaus shipped from the Palmetto Pigeon Plant. All are male—Palmetto has no need for extra males and sells them off cheaply—and they coo and strut in typically male fashion when someone enters the room. One of them, Linus, is a light gray Silver King with a purple iridescent neck. He has been working at the lab for six years to learn and remember images that flash before him on a screen. Cook's lab published these results a few years ago; it was the first time anyone had set an upper bound on the number of items an animal could remember, and the limit was surprisingly large for a thumb-tip-brained creature. Linus can recall about a thousand different images, and the memories last for years.

The laboratory, in an adjacent room, is dark and sounds like the inside of a subway car because of the white noise generator that keeps the pigeons from getting distracted while they work. (They are here to do a job after all, and Cook often praises them as good workers.) The room is filled with "operant chambers," which Cook said is a bad term but preferable to the familiar name "Skinner box." One of Skinner's lasting legacies was his apparatus, which works as beautifully for cognitive studies as it did for behavioral ones. The black boxes, big enough to accommodate a pigeon with a little room to move, are now updated with a touch screen rather than mechanical keys. Behind the screen is a computer monitor that fills an entire wall of the box. Below is a well, or hopper, where pellets of food appear whenever the pigeon makes a choice predetermined as correct. Some of the boxes have cameras wired to small monitors so the researcher can watch the subjects, but in others the pigeon is simply shut inside the box and the scientists see only the stream of data that pours out of it.

When I visited the lab, each of the chambers was set up for a different experiment. One was a memory test, in which the pigeon sees a square in a certain color and then must choose that color from a set of three squares that appeared next on the screen. Later, the rules would change, and the pigeon was not supposed to choose the color it just saw but the next color in a three-color sequence. Cook's team wanted to test how the birds respond to changing rules. They not only seem to adapt to the new rule, but, when the test is run again, their performance actually declines partway through as they start to anticipate the new rule. Other tests were purely visual, trying to determine if pigeons can see the same things we can or are tripped up by optical illusions. Some tested whether a pigeon can perform cognitive tasks; if shown an object made up of different parts, for instance, can the pigeon say how many parts it contains?

That afternoon, a White Carneau called Darlin' was at work. In another room, a graduate student named Hara Rosen was looking through a file of data where Darlin's pecks were recorded. Above the operant chamber, a monitor showed us what the pigeon was seeing: a computer-generated image of a shaded cone protruding from a flat surface. Cook explained that I was able to tell the cone was pointing upward because of the shading, and they wanted to test whether pigeons could also distinguish shapes pointing up or down based just on shading. So far, the bird seemed to be managing it. Another image appeared, a depression in the surface. Cook told me to look at it upside down, and when I did, it looked like a cone. The key, he said, was the illumination. "We assume that light comes from above," he explained. "We want to see if pigeons have that assumption too."

Every few months or so, a paper from one of several labs doing these sorts of studies reports that pigeons can do things like distinguish letters of the alphabet or read expressions on human faces, or, in one paper that garnered a flutter of news stories, reliably tell the difference between cubist and impressionist paintings. Scientists still struggle to understand what it means when pigeons and other animals show surprising cognitive abilities. When pigeons recognize faces, are they really seeing a face, or are they just responding to a visual pattern? They can correctly distinguish shades of blue from shades of green, but do they actually posses concepts of blueness and greenness? The goal of all of these studies is not to understand pigeons but to develop overall concepts about how the brain processes information. Pigeons just serve as a convenient subject.

The operant chambers make use of a pigeon's greatest skill—looking and pecking. These birds are foragers, and they spend a good part of their lives pecking at the ground and hunting for food. The experiments just add an additional cognitive wrinkle into the process; the pigeons learn to peck at something that will result in food coming into the hopper. To researchers who study pigeons, foraging and eating is beside the point—they are just convenient behaviors to get the pigeons to make decisions. But to pigeons, food is everything, and they are exquisitely sensitive to changes in food supply. When kept just a little hungry, as these pigeons are, they are avid learners. Most of them pick up the routine of the operant chamber quickly and are content to peck away at their screens as long as the experiment lasts. They work tirelessly, and their patience alone makes them better subjects than humans. Their natural behavior explains why pigeons are so good at tasks that require both visual discrimination and resistance to boredom, like guiding missiles, inspecting drug capsules, searching for people at sea, or just staring at hundreds of images on a screen. Just as humans have turned from hunter-gatherers to shoppers and web surfers, laboratory pigeons have easily transferred their foraging behavior from the wild.

When Cook sees pigeons pecking at food or courting in a park, he knows that they have more sophisticated cognitive tools available than most people would guess. Like these laboratory pigeons, street pigeons must remember past experiences and be able to make decisions based on those experiences. They may even need to form concepts in order to survive. If behaviorism showed that human behavior could be shaped much like that of animals, cognitive science shows that animal minds are much closer to what we think of as human. Outside the lab, pigeons have been studied for their social behavior. Louis Lefebvre, a biologist at McGill University, has studied pigeons in captivity and in the wild to understand how knowledge spreads from one bird to another—some pigeons, he found, act as "discoverers" of new food sources, while others act as "scroungers" that take advantage of these discoveries. Because they are social birds, pigeons are able learn from one another, and information about new food sources spreads throughout a flock.

I noticed a crack in the door of Darlin's operant chamber and, leaning in, I caught a flash of his white head, illuminated by the monitor's glow, jerking up and down. I thought of Skinner's government evaluators as they took the lid off the black box of his perfect machine, only to find a pigeon pecking for corn. Perhaps Darlin' was on the verge of an important discovery about how the brain makes sense of the world but to him, I suspected, this was just lunch.

7

Escape of the Superdoves

Domestication worked out very well for the pigeon. As the birds were carried throughout the world into new territories over years, centuries, millennia, pigeons escaped. Bird by bird, pair by pair, they found their way into the wild. But in this case, the "wild" was not the cliffsides of their ancestors, but the villages, towns, and cities where they had been taken. Perhaps some pigeons strayed out into natural settings. But most of them stayed put. After all, they now had building facades and bridges to substitute for rocky outcroppings. After feeding on human-produced grain for so long, why leave the food source they had come to depend upon?

Since pigeons had always lived rather casually with people, their owners probably paid little notice when they left. One of the few documented cases of pigeons escaping on a large scale was during the French Revolution when angry French peasants destroyed dovecotes of the nobility—the act was meant to protest the unfair rules governing pigeon-keeping but had the unintended effect of freeing the pigeons from tyranny. For the most part, however, the loss of pigeons was too small an event to be detected by the lens of history.

It's difficult to know when such a thing as a street pigeon came into being. In India and the Middle East, pigeons were more established, having lived in and around cities as long as they existed. In Europe and North America, scattered references describe pigeons living free in cities. The bishop at St. Paul's Cathedral in London complained in the fourteenth century that people were breaking windows while throwing stones at pigeons that lived there. An Italian painting from the fifteenth century shows pigeons resting on a narrow ledge in a city scene. A travel book from the mid-nineteenth century mentioned that pigeons were by then a common sight on London buildings. Pigeons were also common in New York. In 1882, the Park Commissioners in the city "politely declined" an offer to populate Central Park with fancy pigeons, as "there are now as many pigeons in Central Park as the department feels justified taking care of." Pigeons had found new employment cleaning the streets of manure from horse-drawn carriages—while also helping themselves to grain from the horses' feed. In 1907 they were visible in Boston. A physician named Charles Townsend wrote at length about the pigeons of Boston in 1915. He saw them sitting under the eaves of a church and along ledges near Boston Common. "Whole rows of birds," he wrote, "may be seen sleeping peacefully in these situations amid the glare of electric lights and the noise of traffic in the streets."

If street pigeons had become a common sight in many parts of the world, it was only in the latter half of the twentieth century that their numbers exploded, a growth that mirrors the surge in human populations and the growth of urban areas. Pigeon populations were linked to human conditions in cities. They declined in number in Europe during the two world wars, and the postwar period brought a pigeon boom along with a baby boom. Pigeons probably also benefited from the vast improvements in farming after the war, which brought grain surpluses in agricultural areas. In the United States, pigeons followed the population westward.

Although free-living, they were still intimately linked to society, and thus sensitive to war, food shortages, and economics. A good example is in Moscow. Pigeons became common there at the turn of the century but dropped off severely during the Russian Revolution. They became common again in the 1920s and '30s, but declined during World War II. Afterward, pigeons made a comeback, and a policy of sanctioned feeding of birds in the 1950s boosted their numbers twelvefold in five years. In other parts of Europe a tradition of feeding pigeons in the streets created pockets of dense populations in squares like Venice's Piazza San Marco and London's Trafalgar Square.

By this time, pigeons had become pests. In Europe, the irritation began quite early. A study on newspaper references to urban wildlife in the small city of Turku, Finland, found that pigeons were persecuted as early as 1910, when a baron suggested to the city council that street pigeons should be controlled, "as they greatly increase dirtiness in the beautiful city." The council offered ten pennies for every pigeon that citizens killed. But the measures against the pigeons led to an outcry, and a conservation society was formed in protest that eventually shut the program down. It was a scenario that foreshadowed the dynamic between pigeons and people for the next century.

Pigeons had also become something incongruous: an animal that seemed comfortable in a world of concrete and steel. As society industrialized, it increasingly viewed nature as separate and sacred. Pigeons, who preferred the constructed world of cities, did not seem part of that nature. They had also left the realm of domestic animals kept for food, labor, or companionship. They no longer fit into convenient categories. Technically, street pigeons are considered feral—formerly domesticated animals that have escaped into the wild. Many animals go feral, including horses, goats, sheep, pigs, and cats. But with the exception of feral city cats, most tend to leave the human domain: They lose their tameness and their association with people. And in most cases, the animals leave their domesticated state only in a limited territory. With pigeons, the phenomenon happened worldwide on a grand scale.

What made these birds successful? How did they so quickly conquer the world? This compelling question, I found, was mostly ignored by scientists. Pigeon censuses were taken, and scattered studies on feral pigeons were conducted—mostly in Europe, where their numbers had long concerned cities. Despite, or perhaps because of their commonness, few scientists wanted to study pigeons outside the laboratory. Like any other discipline, science has hierarchies. The study of nature favors exotic places, rare species, and untainted "nature." Early natural history texts in the United States emphasized native species in the vast wildernesses of the New World; who wanted to think about the immigrants moving into increasingly developed cities and towns?

A man named Richard Johnston, I soon found out, was one of the exceptions. He was considered the de facto pigeon expert in North America. Whenever I spoke with another scientist, I always heard the same thing: Have you read *Feral Pigeons*? Have you talked to Richard Johnston? But I discovered that even Johnston came to study the birds largely by accident.

Johnston was a natural collector from childhood—the house in San Francisco in which he grew up had a large garden that allow him to explore plant and animal life with the ruthless curiosity young scientists often show. He collected caterpillars in the yard and watched them transform into chrysalises and then butterflies; when he was fifteen he took up taxidermy. Later, Johnston went to the University of California at Berkeley to study zoology, where his particular interest was birds.

In his early career, he made a name for himself by focusing on birds that might have seemed too common to be interesting: song sparrows and house sparrows. He moved to the University of Kansas in the 1950s, and during the next decade he embarked, with colleague Robert Selander, on an analysis of the size and color of house sparrows in North America. They recognized that this bothersome invasive species represented an opportunity to study evolution. All sparrows came from a population of birds brought to New York in the 1800s. In little more than a century, they were everywhere, living in very different climates across the continent. This rapid spread from a known source provided a perfect opportunity to study microevolution, the small changes in populations that may eventually lead to species differences. By measuring the shape and size of thousands of sparrow specimens, Johnston and Selander were able to show that in their short tenure in the New World, sparrows had evolved, developing regional differences in body shape. By then, of course, Darwin's ideas of evolution were well accepted by scientists. What was surprising was how quickly the birds had changed.

Birds that are common have a distinct advantage for a scientist. Science is about data, not anecdote. And the more birds you can find, the more data you can acquire, and the better you can trust your results. Studying cheap and plentiful species like sparrows allowed Johnston and Selander to produce meaningful information.

Johnston came to pigeons near the end of his career, prompted by an architectural twist of fate. By then, he was curator of birds at the Natural History Museum of the University of Kansas. His office was on the seventh floor of the museum building, Dyche Hall, one of the showpieces of the campus—an ornamental building in the style of Venetian Romanesque churches. Above the arched doorway and jutting tile roofs and walls of limestone, the seventh floor provided a decorative top to the building in the form of a long ledge capped with small arches and columns. It was a perfect home for pigeons.

He decided to exploit the situation. After all, grants to embark on research projects in the wild are difficult to come by. Even pigeons are not necessarily easy to study in cities, since their homes are often on private properties not amenable to peeping scientists. In this case, however, study subjects were literally under his nose. Johnston began taking notes on the birds; he constructed platforms on the ledge to lure even more. Eventually about forty birds occupied nests on the museum's ledge, and if one was ever carried off by a hawk, another would quickly move into this valuable real estate. The ledge became a kind of outdoor laboratory. Because the birds would return to their established nests, he could band their young and follow the fates of individual birds. He could gather quantitative data: their size and shape, how many eggs they laid, how long fledglings survived. Much of it could be observed from the windows, with occasional trips out onto the ledge himself.

Meanwhile, Johnston made the acquaintance of other researchers, mostly in Europe, who were also working on pigeons. He traveled to Sardinia and to the British Isles, where he saw wild rock pigeons living on the coasts.

Johnston had already established his reputation as an ornithologist, so he could risk taking an interest in pigeons. No one criticized the work, though some colleagues complained that the throng of birds drew bugs that threatened museum specimens, and Johnston eventually had to dismantle his shantytown of platforms. There was also skepticism as to why a museum curator, with access to the impressive specimen collections in Dyche Hall, would choose to study the birds on the ledge. And a few colleagues were surprised that doing so got him as far as it did.

Johnston even convinced Oxford University Press to take on the dubious venture of publishing a scientific book on feral pigeons. Johnston began collecting all the scattered papers that had been written on every aspect of their biology. During the course of his efforts, a Slovakian researcher named Marian Janiga, who had also studied pigeons in depth, suggested combining efforts as coauthors. Together they literally wrote the book on feral pigeons. It wasn't a fancier's guide nor a history of racing nor a textbook on anatomy and behavior of laboratory pigeons. Instead, it treated street pigeons as an entirely separate phenomenon with their own ways and habits.

More than a century after Darwin upended humans' special place in the world, the belief that natural history includes animals mixed up in human affairs still seems a bit heretical. Few others shared his view. "Professional ornithologists as a group consider feral pigeons to be almost un-birds," he told me. "I think this is a very stupid attitude, and I certainly have been rewarded by studying the birds. And yet hardly anybody other than me has done this."

You could take the view that pigeons, as formerly domesticated animals, provided tainted information. Or you could take the view, as Johnston did, that their complicated history only made pigeons more interesting. In size and shape, for instance, feral pigeons follow more idiosyncratic patterns by geography than many other species, probably because each population may have a unique age and history. In places where pigeon racing is popular, for instance, feral pigeons may tend to be larger because they receive a steady influx of new members from homing breeds. These historical nuances makes studying pigeons more complicated.

"I think many people get into ornithology because they're excited about the prospect of getting close to wild nature," Johnston said. "I'm perfectly happy to admit that that was part of my motivations to getting into this area many decades ago. But I also like the opportunity to treat feral pigeons as though they were natural artifacts rather than creations of humans. And anyway, I treated them like they were real birds."

One of Johnston's findings was that feral pigeons could be studied as a separate entity. They don't occupy their own taxonomic category—in ornithology, the designation of a separate variety within a species applies only to "natural" entities, not domestic animals. But Johnston argued that there were differences in their physiology and behavior that distinguished them from both wild and domestic pigeons. In size and shape, they looked more like wild rock pigeons, even though the domestic birds were their closest relatives. That suggests that ferals have lived outside of domestication long enough that their bodies have begun to return to their ancestral proportions. Since at least some populations of ferals probably have been living free for hundreds of years, there has been plenty of time to undo the work of humans.

But feral pigeons are not simply returning to their wild ways. It turns out that during their sojourn under artificial selection, they gained some qualities that give them an advantage in their current lives. "They transcend the capabilities of the wild birds as well as captive birds," Johnston said. As domestic animals, they were selected for an ability to breed, and breeding is something pigeons can do in spades. Although they focus their breeding on warmer months, they have the ability to breed year-round, and Johnston has seen fledglings in the middle of a Kansas winter.

They were also selected for sexual precocity. Species in the pigeon and dove family are known for starting breeding early in life, but domestic pigeons even more so, and feral pigeons seem to have retained this trait. Even before reaching adulthood, pigeons can form pairs, build nests, and start making squabs. These precocious birds may be less successful at domestic duties, but their practice pays off later when they become more successful parents as adults.

"These pairs have an enormous capacity to churn out new versions of themselves," Johnston said. "They keep at it all the time." Because of this astounding success, Johnston and Janiga jokingly refer to feral pigeons as "Superdoves." Rather than bemoan the rise of this abundant bird, they marvel at its abilities.

By the time I talked with him, Johnston had retired, and he spoke with the unhurried manner of someone with nothing to prove. I decided to pay him a visit at his home in Lawrence, Kansas, not far from the university where he spent most of his career. He and his wife live in a Victorian in the old part of the city, on a bumpy brick road surrounded by lush gardens that had grown over as their tenders stayed indoors during a recent heat wave. His wife, Laura, answered the door and had me wait for him in the sitting room. She warned me that he was losing his short-term memory, as he was now nearly eighty. Johnston was wiry and wore large-rimmed glasses. Our conversations were slow and sometimes stilted as I peppered him with questions, and he gave me brief, measured replies. He avoided speculation or metaphor, keeping his language precise.

I wanted to know why he didn't consider pigeons "tainted," since they had spent time in captivity. "I found it added another level of complexity," he answered. The time that pigeons had spent in domestication could have allowed certain characteristics to emerge and other ones to fall away, which affects how they are now in the wild. He also pointed out the practical advantage to studying a ubiquitous bird that isn't protected by any laws; it's easy to get hold of them.

"See, the kinds of studies that I'm talking about, geographic variation in size and color, for instance, are perceived on the basis of rather substantial samples of specimens," he explained, "and a rare bird isn't going to give you substantial samples." Johnston admits that not everyone gets excited about pigeons. "They want to have something more dramatic like a peregrine falcon; that's what gets them excited," he said, "whereas I get excited about sample sizes. I'm interested in finding out what's going on, trying to approach the truth of the situation— trying to find out what's going on and why."

He paused. "I'm just interested in the process of life. At least among these birds, it's a little more approachable."

I asked him what question he had tried to answer.

"There isn't a simple one question that was asked, except in the most general sense: How were the birds living? How do they manage to do it? And that's so general, that doesn't give you the kind of answer you want."

I mentioned that one question that most interested me was: How did they get to be everywhere?

"They're here because we brought them here," he said. "They almost certainly wouldn't have gotten here successfully without us. But they are as successful as they are *given* that we brought them here. It can be explained only through knowledge of a number of variables. There's no one single answer to what makes them so good at what they do. Because if they aren't good at everything, they just might not make it, which applies to any organism."

It came down to the same Darwinian principles that applied to pigeons as much as any other creature.

"All these characteristics that they have no willful control over, that natural selection seizes upon and either amplifies or diminishes. It can even be viewed as punishing this stupid behavior or rewarding this excellent behavior, if it's behavior you're talking about. If it's physiology the same applies. There are literally thousands of variables that are involved in the makeup of an organism, and all of them are subject to selection. All of them are subject to improvement, and all of them are subject to elimination." He paused, widening his eyes for effect. "It's a wonder that anything ever gets done."

Johnston pointed out all the things that a bird has to attend to each moment—finding food, engaging in the rituals of courtship, constructing a nest, raising young. Pigeons were simply one incarnation of the universal processes of life, but one worth understanding. "The thing that impresses me as a biologist, and perhaps as a student of pigeons, is—what is the phrase—the infinite varieties of their way of life, their approach to living."

"Other than their flexibility, is there anything else you admire about pigeons?" I asked.

For the first time, Johnston broke from his careful language. "They're handsome critters," he said. "Oh, they're beautiful birds."

I went with him to visit his old haunt, the Natural History Museum. We paused in front of Dyche Hall to look at the ledge and its rows of columns at the top of the building. Inside, a flight of stairs led to the offices above the museum space, and I felt guilty when Johnston had to make his way up seven flights because the elevator would only go to the top floor with a key, which he no longer had. After several turns up the stairway and a few pauses, we reached a metal door with a small window. It was locked, of course.

He banged on the door. "Should be some people in here at this hour," he murmured.

We peered through the window and knocked again but no one came. A few minutes more of knocking, and we heard footsteps in the stairwell below.

A trim, gray-haired man gave an enthusiastic call.

"Hi! I haven't seen you in ages. And . . . you can't get in."

"No, I can't get in," Johnston said, smiling.

"Isn't this nice, we're so secure, it's just terrible," he said, using his card to open the door. He introduced himself as Norman Slade, the curator of mammals.

"You're doing okay?" he asked Johnston cheerfully.

"Uh, more or less."

"More or . . . well, heh heh, I'm afraid that's the way the fan hit us these days. Go ahead, take a look at the pigeons, I hope they're there."

The offices flanked a main room filled with metal cabinets. Johnston walked into the corner one, where a young man was working at a computer.

"My name is Richard Johnston, and I used to work here," he explained.

"I remember you used to work right here," the younger man answered. "Hey Dick, how are you?"

"Good, good to see you."

Slade brought in a copy of *Feral Pigeons* for him to sign for a colleague. Everyone looked at him as if he were a ghost. It had been ten years since he'd been in this building.

"Good to have you here."

"Well, just for a short while, just wanted to look through the windows at the pigeons."

"Well unfortunately they're still here."

We examined the ledges from the windows.

"Yeah that's it," Johnston said. "They like these as resting spots mostly. In the corners, they can actually use them as nests."

The pigeons were all off foraging for the day, but we were taken to a spot visible from the bathroom window, where a cubbyhole behind one of the pillars was piled high with dung.

"That's a great spot, really out of the wind," Johnston remarked. We returned to the main room.

"Well son of a gun. Dick, if you can sign that book for Tom, I'll give it to him at the Key West meeting down in Midfield next month. Let me get you a pen."

After he had signed the book, Johnston said, "Want to see skeletons?"

We strolled through the cabinets until he located the rows of boxes devoted to *Columba livia*. He slid one out and opened the top, showing a small heap of white bones.

"Where does it come from?"

"Let's see . . . ," He scanned the label. "This one comes from Italy, west of Alghero, Capo Caccia, Sardinia." It is one of the few places where wild pigeons still live. "This is one that I got. The 4,994th specimen that I took."

He passed the box to me and I held it in my hand. It was light, as if it contained nothing at all.

"The bones are so tiny," I said.

"Well, it's enough. It's sufficient for holding all that muscle and feather."

Next we went searching for pigeon skins in another room. We paced up and down the rows of bird species and had trouble finding the right drawer. "Everything's changed," he remarked. We finally found *Columba livia*, but the doors to the cabinets were locked now, and no one was around to open them.

We can call pigeons superdoves because of their outstanding success. But although they are "real" birds, every bit as part of nature as any other animal, their success is shaped by people. A superhero is simply a person who acquires special powers through some transformative event; domestication gave pigeons the ability to become the superpower they are today.

One day I called up Louis Lefebvre, a biologist in Montreal who had studied learning and social behavior in pigeons. More recently, Lefebvre had been studying innovation in birds—how well they learn new tasks and adapt their behavior in order to find food. In addition to studying two wild bird species in Barbados—a grackle and a dove—he had tested street pigeons near his lab. In one experiment, Lefebvre examined whether birds can figure out how to get at seed inside of a transparent Plexiglas box by pulling drawers or removing a lid, a test of how innovative they could be in procuring food in new ways. It was a test that Lefebvre expected feral pigeons to flunk, and he seemed disappointed that they did not.

"The average Columbiform—the average bird of their family, pigeons and doves—is really really dumb at that," he explained. The Barbados doves didn't pass the test. But pigeons were much faster than predicted. "The only conclusion is there's something special about pigeons that is different from other Columbiforms," he said, "and the most plausible thing is that they're feral."

Superdoves indeed: Lefebvre believes that their time in captivity may actually have made pigeons more innovative than their wild brethren. They certainly don't follow the usual pattern of innovation in birds. In a previous study, Lefebvre and his colleagues looked at the relative success of introduced bird species, which they compared to the frequency with which the birds try new foods—a measure of innovation—and their brain size, a measure of intelligence. In general, birds with larger brains tend to try new foods more often, and these birds also prove to be better invaders. Pigeons, with their paltry brain size, were an exception.

The idea that pigeons may have gotten cleverer at finding their way in the world because of domestication is counterintuitive. We usually think of domesticated animals as having lost the wits that wild animals need to stay alive. But with pigeons there's a key difference: the dovecote.

Unlike other domesticated animals, pigeons never relied on humans entirely for their food. People gave pigeons a little food but allowed them to fly in and out freely to find the rest on their own. And since pigeons were good at homing, they could be allowed this freedom and still return to the same nest and the same mate. "We've done something really special with pigeons that we've done with no other animal," Lefebvre said. "We've domesticated them, but we haven't domesticated the foraging behavior." Even sheep and cattle, which also have freedom to roam, are taken to pastures to graze by their caretakers.

At the same time, however, pigeons were selected for an ability to tolerate humans. And this lack of skittishness helps them survive in the crowds of cities. "You've got the best of both worlds in terms of artificial selection," Lefebvre said. Pigeons gained new characteristics, but they never lost their most important survival skill. Though no one can know for sure, Lefebvre believes that the pressures of artificial selection may have even enhanced their capabilities for finding food, which would explain their unexpected success at his test.

Perhaps these superheroes are like an avian version of Frankenstein's monster—creatures shaped by humans to suit our purposes, only to become our nemeses when they escape. The same qualities that made pigeons attractive transform them into nuisances when they are no longer under our control. As domestic animals, they became hardy, disease-resistant breeding machines, able to live in the crowded conditions of a dovecote. Bring those qualities into urban wilds, and you have birds able to tolerate the incredible densities of individuals they can easily produce.

As I talked with Lefebvre, I grew excited thinking of the historical accidents that produced the feral pigeon. But Lefebvre was not excited about them. The success of pigeons in his test did not make him want to understand their special abilities. In fact, when I asked if he had any new studies on pigeons, he told me he had stopped working with them altogether. For him, their uniqueness made them a poor subject. "Pigeons are good at innovation for the wrong reason, because of artificial selection," he said. "It's not interesting; they're a special case."

I realized then what made Johnston so unique in his conception of what is interesting. I had assumed that scientists ignored pigeons for the same reason we all do: Because they're boring. Now I realized that the very thing that made pigeons special and, to my mind, more fascinating also made them less compelling to most scientists. If I had entertained notions that science today had abandoned ideas about artificial being "wrong," I was disappointed. To Johnston, pigeons' unique experience with artificial selection made them intriguingly complex, but it was a complexity that Lefebvre's sort of comparative research didn't welcome. The overall goal of his research is to look at social learning and transmission of knowledge among birds in the wild—to compare different species and make broad conclusions about how their behavior is alike or different. To do that, a scientist needs a species that is not special but representative: not a Superdove, but an Everydove.

"Choosing a pigeon made sense because pigeons were well studied in labs," he said, "and you can usually bring them into a Skinner box and an aviary, but they also live right out in front of your lab in the city. So it's sort of easy to go between lab and city. But I found out eventually that it's a mistake to work on pigeons, because pigeons are not wild. Pigeons are domesticated feral."

The time that pigeons spent with humans introduced a unique complication, a variable that made it more difficult to compare them to other wild birds. "The pigeon will be doing stuff not just because it's a pigeon, and it's a Columbiform, and it's from a family that has a small brain," Lefebvre said. "It will be doing what it does partly because of that very specific regime of domestication that it went through for thousands of years. So, um, that was a mistake." He gave a curt laugh.

I remarked that few scientists seemed interested in studying feral pigeons.

"No, and it's not a good idea," he said. "Because they're not representative of anything. They're a unique case in nature."

8

A Squab Is Born

When Richard Johnston talked about all the little feats an organism must perform in order to succeed in the world, I began to wonder more about the lives of pigeons. In any given city the birds just seem to appear. There they are, strutting from sidewalk to gutter, bowing and pecking. But there was more to living a successful life than this. Somehow, in the city's dark corners, pigeons were reproducing, replacing themselves with a new fleet. Their numbers suggest they are very good at reproduction, but how do they do it? In the mazes of office blocks and brownstones in Boston, how did pigeons beget pigeons generation after generation?

On a blustery morning in early March, with the wind chill far below freezing, I saw one of the first signs of spring in the pigeon world. As I walked along an icy path to the subway station, I heard cooing and saw a pigeon puff up his breast as he steered a female around a piece of thin winter grass, ignoring the strong wind, fixed only on her. During the short period of time when pigeons are not breeding, they all seem alike. Although male pigeons are, on average, larger than females, it is difficult to tell the sex of an individual bird; even pigeon-keepers usually must wait until breeding season comes. Come spring, the birds transform: The males begin to act like males. They elevate their heads and puff their feathers out, and they begin intensely herding or "driving" their chosen females, who can be identified simply by their willingness to put up with it.

Most monogamous birds get together only for the length of a breeding season; when the next year comes around, they find someone new. Pigeons, however, are among the few bird species that mate for life. Given that they are entering into a long-term bond, choosing a proper mate is critical.

As with many other species, the male pigeon advertises himself and the female chooses. *Feral Pigeons* lists several criteria that females use in picking their mates. Experienced birds are better than very young ones, but birds older than seven years are considered over the hill. If a male has had a previous partner who died—which happens regularly since males tend to live longer—they are more prized as mates. Socially dominant males do better, and pigeons of similar sizes seem to pair up more frequently. And of course, a male whose feathers are scruffy with parasites is passed over for better-groomed companions.

Females also seem to choose their mates based on color, although it's not clear exactly what they're looking for. It's a behavior unique to pigeons that are feral, but it's understandable why such choosiness would evolve. Feral pigeons are probably the most differently colored and patterned free-living bird in the world. This diversity is certainly a remnant from the genetic mutations the birds acquired during domestication.

A typical wild rock pigeon has a blue-gray body with a cuff of iridescent feathers around the neck. The tail is banded in black at its tip, and the wing feathers have markings that create two black bars along the folded wing; the lower back is either gray or white. This pattern, called blue bar, is also found in feral pigeons. But more commonly they have darker plumage that is flecked or checkered, or sometimes covers the whole body in near blackness. And a minority of feral pigeons are dappled in shades of red, brown, and white; occasionally they carry some of the odder color mutations seen in fancy pigeons. The different colorations may affect more than just looks. Studies have found that blue bars may be better at defending their nests; they seem to establish larger territories and protect their young more aggressively. Blue bars are more common in rural areas and low latitudes, while darker birds are common in cities and higher latitudes. But in truth, no one knows exactly what shapes the color diversity of pigeons. I have no idea why the colonies near my apartment were mostly blue and black, while a certain corner of the city a mile away is regularly filled with pigeons that are red, brown, and white.

Like other species of pigeons and doves, *Columba livia* has its own particular ritual of courtship. Before the pair ever get together, the male establishes a territory where the nest will be, and then he shows his availability by cooing. When another pigeon approaches, he will move aggressively—a submissive response from the other bird identifies her as a female. The male dances about her, serenading her with a song that is unique to rock pigeons: coo, roo-c'too-coo. He puffs up like a Pouter and bows his head and occasionally fans his tail as he struts about.

If the female seems interested, they move on to a more interactive phase. The male begins to poke her head and neck gently with his beak, as if preening her feathers. She returns the gesture. He might also show her the future nest she would occupy if she became his mate. This casual dating might unfold over the course of days. Once a pair has agreed on one another, the courtship becomes more involved. She lightly pecks at his beak while lifting her wings, mimicking the begging of a squab. He responds by regurgitating a little something from his crop into her mouth, perhaps showing what a good provider he is. It takes a week or more of courtship before pigeons ever have sex.

Several days after first spotting the courting pigeons on the path, I visited Blackstone Square Park and saw that a pair of pigeons on the municipal building were already consummating their relationships. After all that build-up, it's a rather short affair. She flattens her body a bit with her tail up, he hops on, there's a second or two of flurrying feathers. I've read that females close their eyes during the process, but I've never intruded closely enough to see for myself. After an encounter on the building, the female preened herself while the male burst off the ledge and circled the park, a sort of clapping victory lap.

As spring unfurled, all over my neighborhood I saw pairs of pigeons quietly tucking themselves into the nooks of buildings to breed. Like most people, I had no idea where pigeons nested and how they raised their young, and where all the baby pigeons were. Now I began to figure it out. I started to pay attention to where pigeons flew, and to look for any small sheltered areas where a nest might be.

Though the high cliff faces of building facades might be perfect for roosting and resting, raising babies demands a more secluded spot. Pigeons' homes are as varied as our buildings are, and it took me some weeks of compulsively scanning buildings to notice nests. I saw one tucked into a gap in some copper sheeting along the corner of an apartment building. I watched two pigeons flutter up into the roof of a subway station and disappear into the space left by a missing ceiling panel; inside I heard shuffling wings, rhythmic cooing, and the telltale wheezy squeak of a baby. I found a quiet side street in posh Back Bay where ledges below the windows of the brownstones were lined with spike-laden wire, but the job was bad enough that it left a corridor between the spikes and the wall, perfect for pigeon nests.

In an alley in my neighborhood, I found a larger group of pigeons living on the side face of a yellowish brick building for the New England Conservatory, which overlooked a drive-through alley and small parking lot. Most of the windows of the four-story building held boxy air conditioning units, one of the most useful modern inventions for pigeons. The units either cantilevered over the window sills or were mounted in wood panels braced against the top of the windows, creating a variety of sites. The four-story building, with large windows on each floor, formed a grid of nests where several pigeons could live at any time. I dubbed it the Pigeon Condo.

I visited the spot one bright spring day around lunchtime, which happens to be the favored time of day for pigeons to build nests. Nest-building is a ritual that begins a few days after a pair has mated—at the

Condo, at least four or five new couples were getting their homes ready. The females stayed tucked under the air conditioners, observing the whole affair, while the males gathered materials from the alley below: mostly dry pine needles and some twigs from a nearby tree.

One pair seemed to be experts. Their nest was at the top floor of the building, and the male would drop from the ledge and glide steeply into the alley, pick up something with his beak, and with a strong clap of the wings deliver the goods straight up to the fourth floor to his mate. She would take what he offered and tuck it around her, sometimes needing his help to wrestle a long twig into position. They seemed to be doing better than a pair down on the second floor; the male insisted on hefting a wooden chopstick up to his mate, who gamely tried to place it next to her. Eventually she would move, though, and the chopstick would clatter down to the gutter below, and the male would fetch the prize again.

But even the more expert pair was not creating any work of art. Pigeons are notoriously sloppy nesters and they engage in none of the intricate weaving some bird species use to fashion sturdy nests. They keep the avian equivalent of college dorm rooms; their homes are slipshod affairs dashed together with whatever materials are available, including roots, twigs, eggshells, feathers, string, paper—they have even been known to toss the mummified bodies of their dead young into the mix. An abandoned nest once found at a chemical plant in Michigan was made of hundreds of pieces of metal wire, some wooden splinters, and a pipe cleaner. The advantage of being unceremonious about their homebuilding is flexibility; they manage to make do with whatever is around them, with just a few inches of sheltered space.

Another important nest material is dung. Many bird species keep their nests immaculate, and will remove any dung that their young make. Not so with pigeons. The adult pigeons will often stand and poop outward from the nest, usually onto the street below. But the dung of babies just accumulates around the sides of the nests, building up walls around it. To pigeons, the dung is like adobe and makes a wonderful nesting material. When they can, they'll keep reusing the nest and let it pile several inches high.

The community of pigeons at the Condo spent much of their time in the alley and small parking area in front of the nests. The alley branched from a sloping side street. There was a patch of dirt with a few small pine trees next to it where they would gather and feed—sometimes just on tree seeds but often on offerings from people—a scattering of corn one day, the ends of a bread loaf or a torn bagel another. In the late afternoons, many of them would fly down the opposite end of the alley in the direction of Symphony Hall, a popular gathering place. Below their nests, some potholes in the pavement often held dirty puddles where they could drink. In windy weather they huddled in the lee of the air conditioners and window sills.

I usually stood on the uphill side of the dirt, where I had a better view of the ledges, sometimes standing on a low concrete wall over a roaring air vent from the adjacent apartment building. The alley was not a place for people to linger. Some days, trash bags sat in the dumpsters in the small lot below my perch, filling the whole alley with the smell of sour rot. Cars angled in and out, sending any pigeons onto the ledges; the occasional bicyclist or pedestrian passed through on the way to somewhere. In such an unattractive place, the birds could live in peace and obscurity.

Would their wild ancestors from windswept cliffs see this lifestyle and be mortified by the degradation? It was not the idealized life of a wild bird, but somehow it worked.

Despite all their work to build their nests, they were not always fruitful; some abandoned the nests and moved on. But occasionally a bird would begin sitting, day and night, and I would know that something was happening. Pigeons almost always lay two eggs at a time, one following the other by the length of a day or so. As with many aspects of their lives, pigeon couples follow a precise routine when incubating the eggs. The males take the day shift, from mid-morning to late afternoon; the rest of the day and night is the job of the females. In their first days of the chicks' lives, both parents feed them by producing crop milk, a cheesy substance secreted from the walls of the intestinal pouch. Each parent is able to make enough crop milk for one of the babies for several days, and because of this ability pigeons do not have to spend these days hunting down protein-rich foods to feed their chicks.

Now that pigeons have left their protective dovecotes and lofts, they face all sorts of perils raising their young—among Richard Johnston's Kansas pigeons, an egg had merely a one in three chance of growing to independence. If one of the parents disappears, the entire clutch could be abandoned. Once hatched, the chicks must be incubated continuously to keep their bodies warm—they rely completely on the physical protection of their parents. That's why, even on an exposed air conditioner, younger chicks are difficult to spot.

Indeed, when I visited the alley one day, I saw two well-developed squabs sitting on top of the air conditioner. By the time young pigeons become visible like this, a casual onlooker would mistake them for adults—hence the mystery of why no one sees baby pigeons. But even from my perch on the uphill slope, I could see they were youngsters. Their beaks were too big for their small bodies, and were not tipped with the characteristic nub of white, called a cere, that adult pigeons have. One was small and black, and the other, a bit larger, was a blue bar.

Though they looked like real pigeons, they were still too young to leave the nest and fend for themselves. And so the restless squabs had little to do but sit and look around and occasionally stretch their legs and wings and preen their feathers, waiting to be fed. One of the parents, a blue check with an iridescent neck, was hunting for food in the patch of dirt nearby, while the other, a blue bar, perched on a ledge just above the squabs. If one of the parents landed on the air conditioner, the squabs accosted it, squeaking and flapping their wings to beg for food. If they had not eaten, the parents simply ignored the nuisance. Squabs will squeak and wheeze to beg until their voices "break" at nearly two months of age, and they begin to make low coos.

One day I found that the larger blue bar squab was sitting at its usual place on the air conditioner, but the small black one was shuffling around on the ledge below. It seemed too small to have left the nest. I left the alley for about twenty minutes and when I returned, dinner had just arrived. Crumbled bread and a scattering of seeds had been dumped in the dirt.

After a bout of pecking, one of the parents returned to the air conditioner and, after the blue bar squab chased it about, flapping and squealing, the parent managed to calm the youngster enough to put its beak inside the squab's and pump vigorously a few times, regurgitating a meal. The little black squab on the ledge was craning its neck and squawking for its share, so the parent then hopped down to the ledge and, shuffling to find some solid footing, leaned over and performed its delivery again.

After that day I never saw the little black squab again. Perhaps it fell to the ground or was taken by a hawk. It's common for the second baby in a clutch to be weaker. Whereas the offspring of some bird species are relatively equal, with pigeons the first baby in a clutch of eggs has a major advantage over the second one, which tends to be born small. The second egg might actually function as a kind of "bonus" baby—if food is abundant, its growth quickly catches up to its sibling and both survive, but if times are bad it dies off, leaving the firstborn with a better chance of succeeding. In fact, singleton chicks, which get fed more and wait longer to fledge, have a survival advantage.

The blue bar squab remained in its nest, and over the next days I watched it living on the air conditioner, sometimes nestled between its parents, but more often sitting by itself, preening or lifting its wings in the air as if getting ready for flight. When its parents visited, the now rather overgrown baby became more aggressive in its demands, poking them for its meal. Even as it grew improbably big, the parents continued to feed it, and it remained at its perch. The extended adolescence that pigeons enjoy is a trade-off. It taxes the parents but it also ensures that the birds have time to grow and develop before facing the world on their own.

One day, the air conditioner was empty. I looked at the cluster of pigeons pecking in the dirt and saw the familiar small body of the squab poking gingerly with them, its spindly gray toes shifting about.

I was excited to realize the squab had successfully fledged. Yes, these were just pigeons, squatting among windows of a dirty alley. But this was the stuff of nature documentaries—birth, growth, the cycles of life. And it was happening here, in a small corner of the city far from anything typically considered nature. It all took place on a few square inches of metal, sheltered from the elements by a slight lip of concrete. That life could emerge from the noise and filth of an alley seemed, in its own way, miraculous. I could see what Johnston meant when he marveled that anything in nature got done.

School had begun, and the rooms of the New England Conservatory filled with horn arpeggios and faint piano concertos. The alley began to collect a fine coat of white downy feathers as the birds above began their annual molt. Few people noticed or cared about the spattering of dung that began to cake the sidewalk below the Condo. Over the next weeks, the squab took its place in the little colony. One day it was watching another pigeon drink in the pothole-puddle, trying to balance on a crushed paper cup sitting like an island in the black water. It followed other birds as they perched on the shallow ledges of the building, learning to fly down eagerly when anyone, including me, stopped under the trees where they were sometimes fed. It still begged to its parents once in a while—squabs will keep trying to get food until their parents finally ignore them and they are forced to fend for themselves. But the bird also acted independently, sometimes following the other birds as they flew down the alley while the parents lingered at their nest.

Although it was still a thin little bird, its feet began to redden and the small whitish cere began forming above the beak. Soon it would lose its juvenile features and I feared there would be nothing to help me distinguish it from any other blue bar in the city. It might continue to live in this little colony among the air conditioners or, in a month or two, it might fall into company with another flock and make its home elsewhere in the city, to carry out the business of making pigeons anew.

After spending months skulking in alleys and loitering on street corners to watch pigeons, I took an opportunity to see how the other half lived—the birds that people actually cared about. The ones I could watch with a pair of binoculars without feeling ridiculous: peregrine falcons.

While my pigeon families led a quiet, obscure life on their air conditioners and ledges, the birth of peregrines is a noted event. The peregrine population has been building its way up from near-extinction in the United States since nearly succumbing to DDT in the 1960s. There are still only about fifteen pairs in the state of Massachusetts, and because of their rarity, the state Division of Fisheries and Wildlife (MassWildlife) bands every chick and carefully monitors the birds' status. Three or four pairs live in the Boston area, including a nest at the top of a high building about a block from Pigeon Condo. But to get a closer look at the birds I traveled to Worcester, a small city about an hour's drive west of Boston, where I was told the annual banding of the chicks would be accessible to the media.

It began as a modest turnout—I stood at the top of a parking garage on a weekday morning with a couple of TV crews for the local news, a reporter from the Worcester newspaper, a few mothers with children, and a young man in a tank top who seemed to be hanging around in the hopes that he would be interviewed for TV. Marian Nelson, a public affairs representative for MassWildlife wearing baggy nylon pants and a green polo shirt, pointed out the peregrine nest near the top of a tall office building across the street. It was a large wooden nest box—unlike pigeons, peregrines get specially designed nests once they choose a spot they like, and most building owners are happy to house them.

But like pigeons, peregrines have largely chosen to make their nests in urban areas. Only three pairs in the state have taken up nests outside cities. Peregrines are also cliff dwellers, but they prefer the heights. Many of them now live on skyscrapers and tall office buildings—one pair in Boston even nests at the top of a crane at Logan Airport. Even for a species that falls easily into our concept of "wild," cities can be attractive homes. The peregrines' nests are safer from predators here. Cities are warmer in the winter than the surrounding countryside, so they are buffered from harsh weather. The biggest threat to them is the buildings themselves, which they sometimes fly into.

Nelson explained that these falcons had been living in Worcester for about nine years, though only recently at this site. Last year the male of the pair collided with a building and had to be sent off for rehabilitation for his injuries. In the intervening year, somehow an unpaired male from New Hampshire was able to take advantage of the situation, flying down to Worcester to mate with the female.

We heard a sharp call in the air and watched as a falcon soared overhead, its wings like two crescent swords, and then came to rest along the corner edge of a tall glass building next door. "You can tell them by their pointy wings," Nelson said, and for comparison pointed to her gold earrings shaped into a pair of red-tailed hawks in flight, their wings blunter and rounded.

After several minutes, I could see through a pair of binoculars a group of four or five people in green polo shirts emerge on the roof. One man crouched down and held a net with a long pole between the back of the nest box and the edge of the building, in case a chick panicked and tried to bolt. Another stood waving his baseball hat in the air—his job was to fend off the parents, who began to dive and screech at the group. A third crouched down and reached into the side of the box and smuggled the babies out of the nest.

We left our vantage point, crossed the street, and waited in the lobby of the building, which held the offices of a large HMO. Bill Davis, MassWildlife's district manager, emerged from the elevator, hefting a white canvas bag close to his belly. He opened the handles to reveal three falcon chicks, covered in white down, their large black eyes staring up at the humans surrounding them, beaks agape. Though adorable, they were certainly the biggest chicks I had ever seen. Under the bulb flashes and the bright glow of the video cameras, the biologists began to band their legs. A young woman plucked each one from the bag in turn and placed it on her lap. Two were females and the other male. The females squawked, gray beaks wide, but the male just sat in her lap during his turn, looking out at the incomprehensible scene. Davis affixed federal and state bands to each of the birds, slipping the metal loops over their spiny white legs, and then clamping them down with a bolt, a movement that made each chick cry out and wave its long white toes, looking at once delicate and fierce.

By now, our little crowd had expanded. The sunny, marbled lobby had gradually filled with HMO workers passing through on their way to

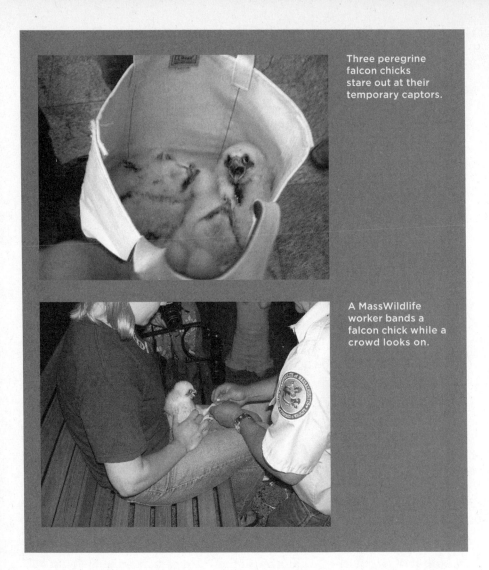

Three peregrine falcon chicks stare out at their temporary captors.

A MassWildlife worker bands a falcon chick while a crowd looks on.

meetings. Each elevator load brought a few more people who lingered, curious about the crowd, mostly women eager to get a look at the chicks and pet them. One of the biologists took a chick in her arms and began circulating through the crowd, letting the women touch the white fuzz and coo over its round black eyes. "Forget about the meeting, this is great," drawled a woman to her friends. If the chick's muscles and talons weren't so underdeveloped I'm sure it would have torn its adoring fans apart.

For a few minutes, even I felt disoriented, amid the heat, the camera flashes, the white down in the air. A video camera bore down on the chick, as the news reporter interviewed the biologist under the camera's bright spotlight. Perhaps the obscurity of alley life was preferable to this hubbub.

The reason I was interested in peregrines, apart from this study in contrasts, was the view they provided into a facet of pigeon-life that gets forgotten in cities—their status as prey. These chicks, though cute and fluffy, were born pigeon assassins.

Pigeons and peregrines are ancient enemies; they have evolved together as predator and prey. Peregrines are not large birds—the female is a bit larger than a crow, the male a bit smaller—but, as the fastest birds on Earth, they are fierce predators. They prefer to hunt from above, diving upon their prey at speeds of up to 200 miles per hour. One of the reasons pigeons may be such acute homers is that in their natural commuter lifestyle they speed home to avoid peregrines watching them from above. Given that situation, what bird wouldn't want to return to its sheltered nest as quickly as possible?

Tom French, who leads the peregrine recovery program in Massachusetts, often gets calls from people who want him to loose a peregrine on the pigeons living on their buildings. You would think the return of these predators would pose an additional threat to pigeons in the city, but in fact the amount of damage peregrines can do is negligible. The birds establish huge territories—in the greater Boston area there aren't more than three or four pairs—so they can't possibly wipe out that many pigeons. Besides, French told me, although Boston's peregrines do eat pigeons, they don't feed exclusively on them. In fact, they often prefer to prey on other species when possible. It's true that peregrines have evolved to hunt pigeons, but pigeons have also evolved to evade peregrines. They are a much more difficult catch than a bird that doesn't have their level of flight control. Peregrines often choose to snack on small migratory birds that fly high and purposefully, rather than pigeons winging through a maze of buildings.

Walking across the lobby, Marian Nelson signaled for me to come closer. She showed me a plastic bag in which the team had gathered some of the carnage surrounding the nest, so they could track what the birds are eating. At the top of the pile of flesh, bones, and feathers, there was the mangled gray carcass of a pigeon.

9

The Urban Habitat

It's a hazy afternoon in Boston, and the intersection of Massachusetts Avenue and Huntington Avenue is crowded with a glut of pedestrians, cars, and buses. Above the roar, unnoticed, a colony of pigeons perches in the filtered sunlight.

The birds that once graced the crags of perilous cliffs now rest atop street lights and line the sloping bars of the traffic signals. They seem unperturbed by the activity below, as if they were still presiding over the sublime vastness of the Mediterranean and not the din of rush hour. And just as they pay little notice to the noise below, people passing by never pause to wonder at the sight of tens and sometimes hundreds of pigeons suspended above them.

I pass this intersection nearly every day, and almost always it is populated by rows of pigeons. They sit motionless, spaced evenly along the bars with eerie precision, a common social habit of perching pigeons. They rest here and, in good weather, bask in the sun. This relaxation is a daily ritual. It gives the birds time to digest the food collected earlier and stored in their crops. Pigeons are social, but not compulsively so. They sometimes spend time alone or in pairs, but often prefer to organize themselves in flocks. Congregated like this, they can better evade a preying hawk or falcon. If one bird perceives a threat it will take off abruptly, and the whole flock will clatter into the air as if moving with one mind. But here, with no dangers around, the pigeons come and go as they please, signaling their intention to leave before taking flight, so as not to disturb the lineup.

Why does a bird choose to spend its leisure time above a noisy street with few trees or patches of grass in sight? And of all the streets in Boston, why do they gather here, time and again, at the intersection of Huntington and Mass avenues, two of the city's busiest thoroughfares?

One clue is in the surrounding buildings. There is a popular vista of cliffs in Arches National Park in southern Utah which has the incongruous title of Park Avenue. From the main road that cuts through the barren landscape, you look down a wide valley flanked on either side by a row of high towers and spires of red sandstone. When early travelers came to that place and saw a strange new topography, they named it after a city street familiar to them. In that same spirit, the pigeons in Boston, New York, or any other city perceive the endless rows of buildings as something familiar: cliffs.

We often assume that what is most "natural" is the best habitat for an animal—a forest is better than a golf course is better than a paved street. But pigeons defy that expectation. They seem just as happy in a parking lot as in a park. To understand why they live in cities, you have to look at things from a pigeon's perspective. Cities fulfill the same needs for pigeons as they do for people: easy access to food and shelter. There was no guarantee that escaping domestic pigeons would do so well on their own; it's a happy accident that our needs match so closely.

Humans have populated their cities and towns with structures that mimic, in all the important ways, cliffs that wild rock pigeons call home. Male pigeons are the ones to choose nest sites, and in man-made environments they are attracted to anything that approximates a crevice or cave—an open window, a porch, an airshaft, a sheltered ledge or pediment. Whether it is natural or not is inconsequential. Feral pigeons can and will find more natural habitats like quarries and cliffs if they are convenient.

In nature, species often have certain niches that suit them best. Man-made habitats also have nuances; not all city features are good for pigeons. Part of the attraction of the intersection of Huntington and Mass avenues is most certainly the buildings. The pigeons cluster not on the tall apartment towers at one side of Huntington, but on two buildings from the turn of the twentieth century at the opposite side. Symphony Hall, a squarish brick edifice built in 1900, has a shallow portico with columns topped with ledges. Across the street is Horticultural Hall, the former home of the Boston Horticultural Society: a brick building from a year later, with a concrete frieze along the top embellished with a series of crescent-shaped vegetal reliefs that look like stone hammocks, each of which can hold several lounging pigeons.

For pigeons, fashions in architecture deeply affect their habitat. Older buildings like these, with their ledges and hollows, generally make better homes and resting spots. You can see the effect in a pigeon census conducted in Milan, Italy, in the 1990s. Pigeons living in that busy metropolis face a varied landscape. In the city center are the older buildings—the massive stone facade of the Stazione Centrale, the scattered nineteenth-century mansion, apartment houses of brick with red-tile roofs, the decorative Teatro della Scala, the Gothic spires of the Duomo and its piazza, where tourists daily gather to feed pigeons and take their picture. As the birds move away from the center, more of the buildings are new—sleeker postwar constructions of concrete and glass with fewer nooks in which to fashion a nest. The census estimated that a little more than 100,000 pigeons populate Milan. The flocks are not evenly distributed; instead, they cluster in areas with a higher proportion of buildings built prior to 1936. Prewar construction makes a better environment for pigeons.

Fortunately for pigeons, we tend to admire and protect older buildings. In a city of modern glass walls, they would have a harder time. Indeed, more new buildings are being designed specifically to thwart the birds, with sloping sills that won't hold a nest and no horizontal surfaces wider than an inch or two. But newer buildings still offer some hidden shelters: The airshaft, a common feature of modern apartment blocks, makes a nice cave for nesting, as many landlords have discovered. It's hard to imagine that cities will ever become entirely pigeon-proof, as the birds have a way of finding a few inches of nesting space in the most surprising places.

When Louis Lefebvre studied the relative success of different introduced bird species, pigeons seemed to be the exception, with their small brains. Though they may have a reputation for eating anything, technically they have a very limited diet of grain and seeds. Lefebvre believes their success is not because they can eat many different foods, but because their chosen foods happen to match ours very nicely. It's easy to see how pigeons would have wanted to hang around human towns as soon as agriculture began. And in many areas, agriculture is still their main source of food.

Within cities, however, they can still find plenty of grain. They can intercept it as it is being transported in ports or held in grain elevators and silos. Or they can find grain transformed into the bread in our trash cans and littering our streets and parks. Much of our food—especially cheap food likely to get tossed in the trash—is grain-based. So a key strategy for many pigeons is to frequent places where people are eating cheap food and throwing it out—or handing a little bit to the birds.

One day, as I approached the corner of Huntington and Mass avenues I saw the entire coterie of birds swirling through the air. At the eye of this tornado was a thin, elderly woman in a green raincoat. At first it seemed like a scene out of *The Birds*, but then I realized they were spinning around her in anticipation. She crossed the first set of lanes of Huntington and approached a rounded strip of sidewalk formed by a turning lane that ran against the main traffic; she reached swiftly into a bag, and with a deft motion dumped its contents on the sidewalk, then crossed the next set of lanes, turned, and walked up the street, leaving the agitated mass of pigeons to their meal.

It was not just the buildings that attracted the birds here. It was a food supply. Over the next weeks, I noticed that two more people frequently dumped large amounts of grain on the same slip of sidewalk.

One frigid and windy January afternoon, I approached the corner opposite Horticultural Hall and saw that I had just missed feeding time: A man that I had seen before was walking toward me, an empty shopping bag in his hand, and the birds were clustered on their feeding ground, feverishly pecking. The food, whatever it was, was almost gone when I reached the corner. I watched them slowly trickle back to perches on the buildings, while about twenty or so stayed behind to peck at the dusty remains of the meal. It was a little after 4pm, and the winter sky was already dusky and pink. Eventually, all the birds gathered again on the face of the hall, squeezing themselves into their stone hammocks and shifting to find room along the ledge below. There was much flapping and jostling as some birds were unable to find any spots they could slide into without getting winged by pigeons on either side. Finally they settled into what I thought was their rest for the night.

But then, at 4:35, the woman came. She whisked past me and removed a package from under her arm—a large plastic bag of birdseed. This time the birds did not see her until she reached the feeding ground, but they soon roused themselves and poured down to the sidewalk. And as quickly as she had dumped her gift, she was gone in the crowd.

Two dinners, less than thirty minutes apart. No wonder these birds were willing to endure the crowded tenement of the building's face, why they massed at this corner. It had everything they needed.

Our understanding of how people shape the habitats of species that live near and among us is still nascent. The field of urban ecology has become popular in the past decade, but before that the idea of doing real natural science in a city was something many people never considered. Eric Strauss, who heads the Urban Ecology Institute at Boston College, told me that there's a common assumption in ecology that "real science begins with an airplane ride."

Many scientists would dismiss pigeons as not worthy of study even in the context of urban ecology, because of their unique experience with domestication. But Strauss points out that such divisions are ridiculous.

"Think of how many calories are locked up inside pigeons," he said. "The peregrine falcon living in the clocktower in downtown Boston doesn't consider whether pigeons were domesticated." Our tendency to dismiss them is just a signal of our enduring ability to pretend certain things don't exist, a consequence of the narrow view with which we evaluate the world. If you think about ecology as the relationship between living beings and a given environment, Strauss said, Corvids—ravens and crows—and Columbids—pigeons and doves—are the birds that dominate urban environments by their sheer numbers.

Another way to look at it is in terms of biomass—the gross amount of living "stuff" that each species contributes. To understand the basic nuts and bolts of how ecosytems work, you have to study how molecules move from organic beings to the soil to the air. You have to get down to chemistry and the actual physical reality of life. If we chose to study species based on their sheer biomass, Strauss pointed out, most scientific grants for ornithology in the U.S. would be for starlings, which are estimated to account for more biomass than any other bird species, and pigeons wouldn't be far behind. (Of course, Strauss's work focuses on neither species but on crows, whose crafty intelligence makes the birds much more promising for funding opportunities.)

But even within the field of urban ecology, scientists have been slow to look deep into the inner workings of cities and towns. Instead, they have focused on "remnant habitats," those that experience change from encroaching urbanization. Many bird studies, for instance, have addressed how population numbers of certain bird species vary between wilderness and suburban or urban habitats. The focus, quite naturally, has been on species that are losing ground to urbanization; far fewer studies attempt to understand those that live quite well in cities.

One study that is changing that pattern is a bird study in Baltimore, part of a large and comprehensive urban ecology project in the city. Baltimore and Phoenix became the first urban sites for long-term ecological research in the country, examining all aspects of the urban environment from carbon cycling to species diversity. As part of it, Charles Nilon, an urban ecologist at the University of Missouri, and colleagues are taking a detailed census of Baltimore's birds—an attempt, he told me, "to focus on day-to-day Baltimore," to understand birds that people see in their daily lives in the city and the factors that shape their abundance. The study can take advantage of the wealth of other data being collected in Baltimore to understand how factors in the human-created habitat affect birds who share it.

Pigeons, not surprisingly, are among Baltimore's most abundant birds, and one of those seen at most survey points. In fact, the only places where they don't appear are those areas designated as urban forests and older neighborhoods with lots of trees; as Nilon put it, "every place you don't see forest you see pigeons." But although the average number of pigeons is quite high, there is a great range of abundance across different points. And Nilon's team found that they could do a decent job of predicting where pigeons are found based on other measures. For instance, when they used data from a street-by-street survey of the buildings and trees in different parts of the city, they found that the more buildings that were on a block, the more pigeons. Another predictor was land use—pigeons clustered in areas with a lot of development, particularly high-density housing like row houses and apartment blocks.

Because of the fairly predictable differences in how pigeons are distributed over the city, Nilon is working with other researchers to develop a spatial model of Baltimore's pigeon population. Using information about pigeons from each of the nearly 140 census points, the model can then predict the probability of seeing pigeons anywhere in the city. Information about land use and vegetation can be added in to help refine the predictions. The result is akin to a topographical map; but it can also reveal how changes in the city—a new development project, a park—could potentially affect the entire map.

Studying birds in a city allows researchers to uncover unexpected relationships between the birds and people that share territory. Urban ecologists have begun looking at how socioeconomic factors affect urban species; a study in Vancouver, for instance, found that wealthier neighborhoods tend to have a greater diversity of bird species than poorer neighborhoods. Cultural factors play a large role in shaping pigeons' habitat. In order to live in cities, pigeons depend on a certain level of tolerance from their human neighbors. With pigeons often making headlines as pests, it's easy to overlook how much the birds rely on our acceptance. Any intolerance they face from people is usually localized; a building owner has pigeons trapped or poisoned, or puts up spikes and nets to keep them away. These minor skirmishes do nothing to discourage the general population from growing. The elderly people who drop bags of seed for the birds at Horticultural Hall, or the ones who occasionally dump spare bread for the Pigeon Condo residents, probably have a greater impact on their livelihood than a single landlord scattering poison.

There is also a more benign form of tolerance: The owners and tenants of Horticultural Hall, who are perfectly aware of the daily food dumps, the accumulating birds, and the pigeon-poop on the sidewalk, would rather ignore it than go through the trouble, expense, and potential backlash from bird-lovers of taking action against the birds or their feeders. The pigeon pairs that I saw nesting had all found places that were out of direct visibility. But there are only so many places hidden in a city. Many pigeons take advantage of the places that people don't care about. In less dense areas, pigeons congregate under wide overpasses and bridges and on the tops of warehouse buildings. In the heart of the city, they manage to find the buildings that are abandoned or neglected. On the blocks of brownstones in my neighborhood, pigeons would collect on those that had been least cared for, that were abandoned or under construction.

Pigeons' predilection for the inner city—clustering where people do rather than in places where trees, bushes, and lawns are more abundant, reflects a behavior called synanthropy—literally, living with man. Synanthropes prefer to live around people and their built environment. Pigeons may prefer us and our buildings because of their history of domestication, but not necessarily so. They may have simply become domesticated because of their preference for seeking out human settlements.

The ability to thrive in new types of environments isn't just a matter of cleverness; many of the species that succeed just happen to have niches that overlap with people, like pigeons do. "A lot of species that are synanthropic specialists have adaptations that let them do very well in built environments that mimic their natural environments," Nilon said. Nighthawks, for instance, nest on gravel rooftops, while chimney swifts nest in traditional chimneys, and house sparrows nest under eaves. Purple martins that breed in the eastern U.S. are entirely dependent upon housing created for them by people. Robins flourish in suburbs, where the neatly trimmed lawns make it easier for them to feed on earthworms. Birds as well as other animals, insects, and plants, are able to find niches that happen to coincide with those of humans. In contrast, many of the other species that succeed in cities and suburbs are generalists—they have the ability to live in many kinds of environments or eat a range of foods, so they can make do in human-modified areas as well as in wilderness.

Pigeons received special training in synanthropy through domestication, but other species are adopting pigeon-like strategies in order to thrive in new environments. For instance, pigeons stay put; by not wasting their energy on difficult and dangerous migrations, they have more time to breed, allowing them to produce multiple broods in a season. Canada geese, which once stalwartly crossed the skies of North America every year, now have caught on to this trick. Many populations have eschewed migration in favor of lounging on soccer fields and golf courses, and in consequence these birds that once inspired poetry have become pests in many cities. Other species have adopted tricks beyond those that pigeons use. House sparrows have learned to gather at rest stops along highways to feed on the insects smeared on the fronts of cars. Crows will drop seeds into the line of car traffic to crack them open. Whereas pigeons are fortunate to find food and shelter that matches their needs in cities, other species thrive by being neophilic—they readily try new things and thereby discover new food sources and ways of surviving.

The term synanthropy is not widely used in the United States, and is most commonly used to refer to insects and animal pests, the cases in which this trait is most annoying to us. Perhaps synanthropy is seen as an aberration, a corruption of natural behavior. But given the fact that the earth is becoming increasingly urbanized, synanthropes may have hit upon a winning strategy. Not all species that live in urban environments are synanthropic. Most birds are able to live in cities by relying on the parks, lawns, trees, and shrubbery that pepper the landscape. But even these "natural" features are shaped by humans, as the trimmed lawns illustrate. As humans spread, these subtle features of the environment may have a growing impact on which species win out.

10

Defining Pigeons

I returned several times to Blackstone Square park to watch the large colony of pigeons return daily, in ritual fashion, to their resting places in the corners of a window ledge. One day, I arrived there in darkness just before sunrise, because I wanted to see them wake up.

They were immobile and quiet, dim spots at the top of the building. But soon I caught flashes of white through the darkness as a few birds began to expose the pale underside of a wing. A few minutes later, more stirring, and soon the birds were lifting their wings almost in unison, a slow undulation. Tails popped up, heads bent to preen. The birds began to shift and bump into one another on their crowded ledges. A few flew from the nests to the top of the columns. One pigeon fluttered to the very top of the building and flew off. But though they were all awake now, most of them stayed in place as if waiting. I suspected what they were waiting for, and he arrived a few minutes later: Raymond.

Raymond, a large man with a shaved head, wore headphones over his ears and entered the park slowly, head bowed and feet placed deliberately as if enacting a walking meditation. When he reached an area of the grass opposite the pigeons' building, he reached into a bag holding three loaves of bread. Five birds swooped down to the ground, then a few more flew down to a lower ledge. Raymond began to pull pieces of bread out and toss them out in a semicircle around him. The rest of the birds were hesitant at first, drifting down and then swooping around several times before shuddering to the ground in a ripple of wingbeats.

I had met Raymond on a previous afternoon when I found him feeding the pigeons in the same way. He told me he worked part-time at a substance-abuse counseling center nearby, and that he often stopped here to feed the birds because he liked to observe their behavior—that they help him get through the day. He had named two of the birds—one with a missing right foot called Andy, and a white pigeon with wing-shaped black markings on its back, named Gabriel.

When I first saw Raymond, I felt disappointed and a little annoyed. Here I was, playing naturalist, trying to watch pigeons in the wilderness of the South End, and it turned out they were getting daily meals from Raymond. But in assuming these birds should be off foraging, I had fallen into the same sort of mental trap that keeps people from noticing pigeons in the first place, those imaginary divisions between man and nature. These birds had a history with people that stretched back thousands of years; they lived in our midst. Obviously, their version of wildness includes people. So to understand pigeons it was necessary to understand someone like Raymond.

People have diverse attitudes toward pigeons, and each affects the other's ecology. Wherever pigeons are persecuted, they also win defenders. Fervent ones. When I first became interested in pigeons, I easily located groups of people who could be given the unlikely title of pigeon activists. The most vocal were based in New York City; these were urban rescuers who adopted hurt pigeons and protested against any landlord who tried to trap or kill pigeons on buildings. They posted impassioned defenses of pigeon rights on websites.

Devotion to feral pigeons in New York City is an incongruity. One could argue that of all bird species, pigeons are among the least in need of defending.

D r. Anthony Pilney, a bird specialist at the Animal General veterinary clinic in Manhattan, pulled his patient out of a soft black pet carrier to examine its injury. This bird was a street pigeon, not a pet, but Johanna Clearfield hovered nearby with all the concern of an overprotective owner.

Pilney was young and amiable and dressed casually in a loose plaid shirt. He often saw birds like this, brought in from the street by concerned people. Sometimes it was a person who just happened to see a hurt pigeon and felt compassion, but more often it was a regular like Clearfield who is constantly looking for hurt pigeons, who feeds the birds and acts as a caretaker for neighborhood flocks.

The bird was all white; its tail feathers were caked in greenish excrement from its journey in Clearfield's pet carrier, but its face seemed placid and shy—dovelike, you might say. Of all varieties of feral pigeons, white ones probably receive the least human ill will. White homing pigeons are often released at weddings and other functions as "doves," which gives feral populations an influx of white plumage when the birds fail to return to their lofts.

The pigeon's right foot was folded over and pulled in against its body. Pilney held the foot and examined the curled toes. He said he would have to X-ray the bird to see if its bones were broken. Clearfield asked him tensely if he would contact her in case he deemed the pigeon unreleasable. Unreleasable pigeons are usually euthanized unless a suitable home can be found for them. "I know a woman with a sanctuary upstate," she said. He assured her that he would give her a call.

Later, Pilney told me that few vets specialize in treating birds, and even fewer are willing to see stray wildlife. Both are counts against pigeons. Veterinarians hold different philosophies on whether and how to intervene in the health of wild animals. "I'm personally not a big fan of a lot of intervention. I'm a fan of mother nature," he said. "My issue is, once somebody intervenes, then we have the obligation to follow through." Those who are willing to treat wild birds often find themselves with lots and lots of pigeon patients.

"There is a small subset of society that are extremely interested in the welfare of pigeons," he told me. "They are what we call 'pigeon people.'" Many of these people were retired or simply had the time and inclination to scour the streets for hurt pigeons. It's not that pigeon people fall into a specific type or category, Pilney told me. But there is some distinguishing quality. "If you put a bunch of people in a lineup," he said. "I could pick out which one was going to rescue a pigeon."

Johanna Clearfield is a thin woman in her forties with cropped hair and sharp dark eyes. She sees her pigeon activities as rooted in the tradition of social activism, and often uses the terminology of social justice when she speaks about pigeon rights. In her mind, she is fighting for nothing less than the soul of society. And the very dubiousness of her cause only fuels the activist instinct. "I'm a civil rights person to the nth degree," she told me. "So I don't care if it's not a popular cause, because I believe in right and wrong, period. A lot of people thought it was fine to lynch blacks in the South." She keeps a blog and leads a group called the Urban Wildlife Coalition that is fighting to convince the city to adopt a humane management policy for the city's pigeons.

She also belongs to one of the more organized groups of pigeon lovers in the city, fittingly called Pigeon People. They have a Yahoo! group to which they post pigeon activities in the city, and they hold meetings to debate pigeon policy. Shortly after seeing Clearfield, I went to a third-floor walkup in South Brooklyn for one of the group's monthly meetings.

The chief Pigeon Person is a man named Al Streit. In his fifties, Streit works at a nonprofit; he organizes pigeon activities in his spare time. He neatly encapsulated his thinking for an article in the *New York Daily News*: "Pigeons love people and there's nothing wrong with loving them back."

A short man with a thick Brooklyn accent, glasses, and black hair thinning over his crown, Al Streit had a soft spot for animals in need. When I talked with him before the meeting, Streit told me that most pigeon people can trace their infatuation to one incident when they encountered a sick or injured bird on the street. Even if they never had an interest in the birds before, the sight of a helpless animal touches them. Streit likes to say that one of the most important functions of his group is to help people "come out" as pigeon lovers. It's not easy to admit you care about street pigeons—in some circles it's probably akin to admitting to a mental illness. It takes a special kind of compassion to feel affection for an animal that others ignore or dislike, and a certain boldness to admit your feelings publicly. And it takes something even more to seek a community and become vocal about your feelings, to organize and proselytize—to become a bona fide Pigeon Person.

In the meeting room, six people gathered around a large table. By the meeting's end the group grew to a dozen. But most of the meeting was dominated by three voices: Streit, Clearfield, and Laurie Spiegel, an old college friend of Streit's and an avant-garde music composer, who wore her long gray hair in a loose ponytail and tended to speak in a dry monotone. Also in attendance were Streit's wife, a few quieter pigeon lovers, and a self-proclaimed sparrow man who came to the meeting to gather information about other urban birds.

The meeting was chaos from the beginning. Streit began by reading from a folded piece of paper on which he had typed up the agenda, but the group quickly veered to other topics. Clearfield kept referring us to the ten Xeroxed pages of references she had prepared for the meeting. Laurie slowly intoned objections beneath other people's words, flustering Clearfield. Streit began snapping at everyone, and a few quieter members tried nicely to suggest that Streit was being a tyrant. After about a half an hour, someone pointed out that Streit still hadn't gotten through the list of meeting topics on the agenda. It all began to look like nothing more than a spectacular clash of personalities, but the Pigeon People's trouble was more than its squabbling leaders. Their disagreements were about more than how to run the meeting: They were about our relationship to the birds and their place in nature.

Pigeons are among the few birds not protected by federal law—it is legal to trap, poison, or otherwise kill the birds, though the Humane Society will object if they deem a method to be cruel. They and other groups argue that pigeons are wildlife and should be left alone; feeding pigeons, they say in their brochures, only encourages an unsupportable population. But most Pigeon People do not agree with this stance. Spiegel and Clearfield believed that feeding pigeons is a simple act of humanity for an animal stuck in cities with no natural food source. But they also argued that something must be done about the pigeons; they favored a hands-on, humane plan to reduce their numbers. Streit's sentimentality led him to contradictions; he saw pigeons as outdoor pets that we should feed and love, but when it came to dealing with the consequences, he balked at managing their population. The question of whether we consider pigeons wild or not is central to how people relate to them. How can you organize a group around an animal that no one can define?

A debate erupted over the merits of a possible pigeon-control program, which they suspected New York City might eventually adopt. They discussed a group called PICAS—Pigeon Control Advisory Services—a sort of one-man consulting service run by an Englishman, who used a method of housing birds in lofts and removing their eggs. Spiegel and Clearfield felt that a plan similar to the PICAS approach would be acceptable if it included taking the eggs away, but didn't limit feeding, which Spiegel termed "population control by starvation." At some point in the discussion, Streit announced that he was totally against any form of population control for pigeons. That statement sent Clearfield and Spiegel into fits. We all sat silently as the three of them talked over each other:

"I am absolutely against you because you're against the birds."

"I'm on the side of the birds."

"How could you be against . . ."

"What about sparrows? What about starlings?"

"They have adequate food! They have adequate food . . . in Central America, in South America there's overpopulation . . ."

"Are you gonna kill people off who live in the slums of South America?"

"We're talking about birth control!"

"There's too many babies for them to feed and it keeps them in poverty, a cycle of poverty!"

"The United States allows birth control pills. Al, are you against spaying and neutering dogs and cats?"

Clearfield was getting worked up. "Because I see these poor birds and I see these little mother birds who are so stressed out trying to care for these little baby birds. They have no food source and they have no . . . " she began to shout. "You are stressing out these mother birds, they cannot handle it! They've got two babies coming every four weeks. Would you like to do that if you were a mother?"

"A lot of us would like your explanation . . . "

"Would you let me talk?" Streit snapped.

After another spate of interjections, a rare space of silence descended on the meeting.

"Why am I against population control at all?" Streit began again quietly. "I'm not for population control for any other animal and I don't see why you single out pigeons."

"All veterinarians tell you to spay and neuter, it's humane," Clearfield said. "It's *humane*. These animals are stressed out, they don't have the resources. . . ."

"Cats and dogs are supposed to live in our homes," said Streit. "Pigeons are not supposed to."

"That's ridiculous, there's not agreement here . . . ," Clearfield was saying.

"This is nature. Can't we just leave them?" said Streit.

". . . starvation and cruelty . . . ," Spiegel was muttering.

"How many people here believe in some kind of pigeon management or some kind of population control?" Clearfield demanded, prompting one of the quieter members to plead to table the discussion until they all had a chance to think about it.

After the meeting I had dinner with Streit and his wife, Gela, at a small Mediterranean restaurant nearby. She had been quiet through most of the meeting but when she did speak it was carefully and calmly. Gela works in social services, helping disabled people find jobs and maintain normal lives; she is disabled herself, and her right hand curls in a way that reminded me of the wounded pigeon that Clearfield had rescued. When Gela first met Al, he was living in a basement apartment with thirteen cats. The cats kept showing up on his doorstep, probably because he never turned one away. Gela had never been an animal person, and thought that all cats seemed the same. After about a week with Al, she had learned the distinct personalities of each creature and had developed an attachment to one in particular.

Over the years they lost almost all of their cats but picked up three pigeons, either abandoned babies or injured birds. But keeping pigeons is difficult. The paranoia that pigeon lovers often exhibit is not unfounded. In New York, it's illegal to keep a feral pigeon in your apartment, because pigeons are classified as livestock, an antiquated distinction from the days when their largest role was as a food source. The people who take hurt or sick pigeons into their homes are in fact sheltering fugitives. At the time, Al and Gela were looking for someone to house the birds for them so they wouldn't get evicted.

As we munched on falafel and Greek salads in the small, dark back room, Streit recounted the tortuous history of pigeon activist groups online and in the city. After watching today's clash, I was no longer surprised that the formation and dissolution of various pigeon groups could be fraught with drama. He told me about a rehabber named Fred who started an online group to discuss pigeon rescue but got depressed and began lashing out at members. Eventually he had a falling out with Streit when Streit wanted to take calls about other birds on the hotline they set up. Fred seemed to feel Streit had overreached his bounds, telling him, "Al, the power's gone to your head."

There was Anna Kugelmas, a longtime pigeon advocate whom Clearfield had warned me was truly insane. ("No joke, she's been arrested for various assaults on people," Clearfield said. "She's like the old lady who hits you with an umbrella.") Streit seemed to hate the woman too. He told me that once she had sent out fliers all over town for a meeting of a group called "Pigeon People"—the very name of Streit's group—and that was the last straw for him. "I can never forgive her. Now, Fred I would forgive like that," he said with a rueful snap of his fingers.

At their best, pigeon people show a unique compassion in noticing and caring for the everyday suffering of a disregarded creature. Unlike the large animal societies that concentrate on the rarest and the cutest of creatures, pigeon people draw attention to the everyday and raise it to something of value. They remind us some kind of wildlife, even the plain old pigeon, is better than the alternative: a city with only ourselves for company.

But the shift of perspective that allowed them to value a forgotten bird can easily lead them to lose any sense of scale. The compassion turns to paranoia. The nobility turns to self-righteousness. Loving the unloved can give you the feeling that the whole world is against you. They even turn against each other. (The Pigeon People listserv chatter is filled with little conflicts. "Don't poop on me from your high perch," one woman scolded another in a particularly nasty exchange.) Pigeon people often defend their birds against the moniker "rats of the sky." But who defends the rats? We can't escape the hierarchies we create without elevating everything. At some point the argument always unravels; it's difficult to get behind saving an animal that is one of the least endangered birds on the planet.

If the pigeon people of New York City can continue to organize and advocate without imploding, they may just accomplish something. They may never reach their goal of convincing the residents of the city to love pigeons, or persuade the scattered state agencies to adopt a rational pigeon management plan. But they may at least challenge the city to define what pigeons are and what place they have in our cities. The process of defining the birds may only illuminate their contradictions, but in any case it will make it harder to simply ignore them.

Though pigeon activists may seem extreme in their views, they argue for a sense of responsibility toward our environment that runs deeper than the latest popular cause, and that perspective is worth considering. Johanna Clearfield views pigeons as "grandfathered" here as part of our society. "Pigeons are here, and they shouldn't have been here," she said, but because we brought them here, we have an obligation to look after them.

She has a point. Richard Johnston echoed a similar sentiment in explaining why the historical context of the birds matters, why we shouldn't just abandon responsibility for our manipulations of nature. "In North America, in big cities, pigeons are there in a large part because, one, we brought them there, and two, we modified their behavior," he told me. Their history is linked to our own. "They're really a part of us in a way," he said. "So I'm not sympathetic with people who consider pigeons an odious blot on the landscape. We reap what we sow."

In the summer of 1935, an order was sent around to office workers in the buildings on Broad Street near the New York Stock Exchange to stop feeding flocks of pigeons at the windows of office buildings. A story in the *New York Times* reported on the incident, poking fun at the group of pigeons that had for years been regularly watered and fed by workers, enjoying "special privileges that put them in a pigeon class by themselves." These birds were accustomed to eating chicken feed and having baths. "They no longer mingled with the 'common' pigeon on the street," the article read. "They were aristocratic pigeons."

One disheartened girl who worked at the Chase Bank Building had been feeding a flock of sixty or seventy pigeons that came each day to her office window. She worried that the birds, accustomed to a leisurely lifestyle, no longer knew how to find their own food, and suggested it might be possible to arrange for these "spoiled" pigeons to be fed in a city park. However, in a quick field trip to the financial district, the reporter assured readers that the Wall Street pigeons could be found looking "lonely but not undernourished," while others seemed downright "fat" and "satisfied with life."

The lighthearted article presaged what would come in the next decades, in more serious terms: battles between pigeon haters and defenders, arguments about the ethics of feeding birds, about trapping and poisoning them, about whether they had any place in cities. And the incident seems to touch on a problem that has never really been solved: No one can agree on the proper relationship between pigeons and people in cities. A 1959 article in the *New York Times* alluded to the "bitter war" now being waged in the city between pigeon lovers and pigeon haters (or "pigeonogynists" as the author calls them).

One of New York's Pigeon People feeds a flock near her home.

The office workers on Broad Street were following a human inclination to befriend animals by feeding them. Official literature on pigeons discourages people from feeding pigeons because they are wildlife. And yet a large industry caters to people who want to feed more popular wild birds—including specialized seeds and bird feeders designed to catch the attention of the species deemed attractive. All of it is officially sanctioned; the esteemed Cornell Ornithology Lab, for instance, sells bird-feeding products on its website. So the argument that feeding pigeons is fundamentally wrong seems unfair. The difference, of course, is that pigeons are not aesthetically pleasing like cardinals and bluejays. They insist on living on our buildings rather than amusing us from a distance in the backyard, and they are bigger birds capable of producing large amounts of dung on those buildings. Living with urban wildlife is never easy, but it's particularly hard when we are in direct competition for territory.

People who feed animals are not operating entirely out of generosity, of course. I suspected that Raymond and the pigeon-feeders I'd seen in Boston felt some satisfaction seeing the flocks anticipate their arrival and swarm down to greet them. Feeding establishes a bond but also a sense of control, of being needed. What struck me about the New York pigeon activists is how vehemently they argued that it was not just they who needed the pigeons but the pigeons who needed them.

Johanna Clearfield argues passionately for the need to control pigeon populations to keep them healthier, yet still insists on feeding the birds. She believes that it's an act of humanity to feed birds that have limited food sources.

Certainly, life is tougher in the urban wild than in dovecotes. Most feral pigeons live only a year or two, and a smaller percentage live up to four years or more. In contrast, some domestic pigeons can live into their teens, and a World War I veteran named Kaiser was reported to have lived thirty-two years. But is the quality of life in the city much worse than in any wilderness? I asked Dr. Pilney what he thought about the so-called plight of pigeons in New York that Johanna had told me of. "I think that's what a lot of pigeon people feel," he said. "But you know, I see millions of happy healthy pigeons that live in this city." And despite the occasional injured pigeon he treats, he certainly doesn't see the majority of birds suffering. "I don't think that they have such an awful life," he said. "I can't walk to work three blocks without seeing pigeons eating a bagel, or a slice of pizza, or, sorry to be gross, but a pool of vomit." Pilney feels that most of the feral birds he sees are generally shiny-feathered, well-fed birds, and the emaciated ones that pigeon people occasionally bring in are skinny because they're sick, not because they are hungry. But he admits it's an open question, and a good one: Do we need people intervening to make pigeons' lives better?

When I was sitting with Clearfield in the waiting room of the veterinary clinic, she said she didn't see pigeons as thriving in a city like New York, though most scientists and urban ecology experts I asked felt that pigeons were perfectly adapted to city life. "I wouldn't say pigeons are happy here," she said. On the contrary, she believes the birds are suffering, that they are starving, and living marginal lives. As we talked, a man in tattered clothes passed by the window, prompting a sigh from Clearfield.

"I don't think the city is an attractive environment for anything," she said nodding toward the gray world outside. "I despise the city."

Thoreau wrote that the mass of men live lives of quiet desperation; according to pigeon people, the mass of pigeons suffer a similar fate. From his cabin in the woods near Concord, Thoreau saw the life of modern man as crowded, materialistic, repetitive, laboring—a life of purposeless toil in society cut off from everything natural and wild and spiritual. His writings helped to create a reverence for Nature as something to be sought apart from society. I wondered if some of the despair that pigeon lovers feel comes from viewing the birds through this well-worn lens. If nature is something pristine and sublime, elevated above the daily buzz of the common man, then an animal living in the city seems like a tragic figure, displaced from Eden into a world of banality. Whether the pigeons see it that way is another matter.

11

Pigeon Mothers

I watched the colony of pigeons living at the corner Massachusetts and Huntington avenues swell. In the afternoons they clamored onto the street lights, hovering to find an open position above the street. In the evening, the face of Horticultural Hall was dotted with pigeons, squeezed onto surfaces of the stone frieze like too many currants stuck in a bun. Their numbers rose and fell a bit throughout the day, but none of the birds had much reason to leave this spot for long—ample food showed up two or three times a day on schedule. Even dovecote pigeons never had it so good.

Though the bickering of New York's pigeon lovers may seem marginal, it does have real consequences. Whether we choose to view an animal as a sacred symbol, a laborer, a pest, or a plaything affects its ecology. It is not just a cultural debate; our internal biases help to shape the external conditions that other species face. In the case of my corner, when just two or three people made the decision to feed the pigeons daily, it transformed the intersection into a wealth of resources for the birds. And rather than oppose the development, the tenants and owners of Horticultural Hall, decided it was easier to turn a blind eye. Just a few dedicated caretakers in an environment of indifference allowed the pigeons to flourish.

The potential for people to affect the ecology of pigeons is illustrated by one of the few documented successes of pigeon control: in Basel, Switzerland. The city launched a pigeon-control program several decades ago; in a short span of time, the population of pigeons decreased dramatically, a phenomenon that the city attributes to its scientifically based, ecological approach to pigeon control. What's interesting is that its approach ultimately depended on controlling people rather than pigeons.

Despite a large market for pigeon control services and products, it is very difficult to make a significant dent in the pigeon population of a city or town. The effects of most products and services are local: At best, they can pigeon-proof the facade of a building or briefly free the area of birds by killing them until new ones take their place. A new contraceptive drug for pigeons may be more effective, and several cities are eagerly putting it to use, but its success is still unproven, and one Italian study suggests it might only have a fleeting effect on population size.

It's only possible to truly eliminate pigeons in very special circumstances. In their journey around the world, domestic pigeons inevitably made their way to the Galápagos Islands, which were once a free-for-all for settlement. People, as they've been wont to do since diluvian times, brought animals on ships with them including pigeons and other domesticated species. And the pigeons, as they've been wont to do for just as long, quickly escaped and established feral populations on the islands. Now that the vast majority of land on the islands has been set aside as national parkland for tourists looking for pristine nature, not bastardized livestock, the park authorities are on a mission to eradicate every species of plant, insect, or animal that is not native to the Galápagos. Pigeons were relatively easy, because they were confined to towns and easy to spot. During the course of a few years beginning in 2000, the entire population was killed off.

But without the luxury of an island's isolation and the Galápagos' strict quarantine procedures—not to mention public support for destroying nonnative species—other locales can never keep new pigeons from moving in; they can only hope to reduce the population to a manageable level.

Basel's well-publicized success was devised by a Swiss pigeon researcher named Daniel Haag-Wackernagel. As someone who had focused his research on the ecology of feral pigeons for more than thirty years, he could probably be called the world's foremost expert on pigeons, and has published studies about how urban pigeons live and how humans affect them. But because he was so involved in pigeon control, I had assumed he was one of what Louis Lefebvre has dismissed as "pigeon haters," people who use scientific arguments to get rid of what is essentially an aesthetic problem. They say that pigeons are diseased, that their crowded conditions are dangerous or unhealthy. I was skeptical of such arguments because so many other scientists contradicted it; it seemed like an argument designed to appeal to our images of slums and of cities as harsh environments for animals. Haag's current work, for instance, involved studying parasites living on pigeons and diseases they could spread. But in contrast, his website was filled with photographs of and tributes to the history of this friend to man. He had written a detailed book about their history in culture. No one who waxed so poetical about pigeons could be considered a hater.

Basel sits at Switzerland's northern border. Though it is the country's third-largest city, its population is a modest 500,000, and it is surrounded by rolling agricultural fields. The Rhine cuts a picturesque arc through the city center; on one bank, the old city has narrow, cobblestone streets that wrap over hills, while rails for electric trolleys split the larger roads. Though it has a historic feel, Basel is also home to a busy university and several pharmaceutical companies—it attracts business travelers rather than the Alps-struck tourists that fill other Swiss cities.

I visited Basel during an unseasonably warm spring when the entire city was abloom. The morning after my arrival, I met Haag in his office at the University of Basel. In his mid-fifties, with tan skin, dark hair that parted in movie-star waves, and strong chin, he did not fit the part of a pigeon person. The spacious office was decorated with ancient dove statuary and pigeon artwork. On an upper shelf perched three stuffed examples of *Columba livia*—a wild rock dove, a dovecote pigeon, and a feral pigeon.

Haag had begun working on the pigeon problem while getting his doctorate in the late 1970s and early 1980s. At the time, city officials felt the birds were doing too much damage to buildings and parks in the precious city center. They had already attempted a ban on feeding the birds, which met with protest, and had tried killing tens of thousands of pigeons, to little avail. Finally they decided to do what few cities do: invest serious money, in this case 20,000 Swiss francs, in researching a solution. Haag took on the task and spent five years studying the pigeons of Basel. He looked for parasites in the birds. He dissected their crops to see what they had just eaten. He looked at where they lived and how they raised their young. He then developed a plan for the city, which he presented as his doctoral thesis.

"If you want to find a solution for a city, you have to make a scientific investigation about the status quo," he explained in lilting, accented English from behind his desk. "You must catch one hundred pigeons and look at the health state of them. Are they infested and with what? Which diseases and which parasites? And what do they eat? You can see this from crop analyses. And then if you know this you can build up a strategy. You must find the weakest point on the system."

Haag didn't believe that pigeons could be controlled in Basel with killing. Pigeons have such a powerful breeding ability that killing some only makes room for new ones. They are capable of producing five times their population every year; to even make a dent, you'd have to wipe them out entirely, as in the Galápagos. He also didn't believe it was possible to get rid of pigeons' habitat. They are resourceful enough to build nests just about anywhere. You can cover a building with spikes and nets, but the pigeons can just move next door. It would be expensive—and nearly impossible—to block every possible nesting site in the city. He was also doubtful whether a contraceptive pill would work; it was impossible to treat every pigeon, and if you only reduce the output of a few pigeons, you just create an opening for others to churn out more squabs.

"If you look at the ecological system of the feral pigeon, there is only one weak point," he said. "It's only the food supply that is steering the population size."

It has been debated where feral pigeons get their food—whether they find natural foods or rely on human handouts. Many studies in Europe have found that pigeons feed predominantly on grain from agricultural fields rather than bread or trash, but the latter undoubtedly becomes more important in dense urban environments. Haag found that Basel's pigeons got only a small proportion of their food from wild seeds—he estimated that a mere 10 percent of their food came from natural sources. The rest included some trash or casual feeding, but the vast majority of the human-derived food, he believed, came from a small group of people feeding the pigeons.

These were not people who casually tossed a piece of bread when they could spare it. "There were whole groups of old women who fed the pigeons," he said. "And if one was ill the other went feeding in the same place." In Basel, such a person is called a *Taubenmutter*, or pigeon mother. Some of the pigeon mothers were actually men, but most seemed to fit the profile of an elderly person with little money and few social connections. The stereotypical crazy pigeon lady. And their actions were very similar to what I had seen in Boston and New York.

The concerted feeding of pigeons leads to what Haag terms an "ecopathological state." With their food provided for them, the pigeons no longer have to take trips around the city or out to the countryside to feed. With so much less energy needed to find food, the birds are free to breed with abandon. After living in confined dovecotes, they are able to tolerate the resulting dense crowds more than wild birds would. Haag argues that these crowds are not just unsightly, but actually unhealthy for pigeons.

But most people saw feeding pigeons as a harmless act, even a charitable one. Haag decided that the only way to change the state of affairs was to turn public opinion against feeding. But he wanted to do so not by turning people against pigeons, but by convincing them that feeding the birds was harming them more than helping them.

With another 250,000 Swiss francs from the government, Haag led a public information campaign to try to persuade people to stop feeding pigeons. He and his colleagues argued that people who feed pigeons only encourage their population to swell to unhealthy proportions. "Feeding pigeons is animal cruelty" one sign read, and another: "Protection of animals is: not feeding pigeons!"

The campaign succeeded, but not because everyone became more enlightened and changed accordingly. Haag believes that the success of the campaign came from creating an environment in which feeding was seen as socially unacceptable. In four years the pigeon population dropped from an estimated 24,000 birds to 8,000. There was never a ban on feeding, but Haag said that many of the regulars "didn't dare feed" because they were looked down on and verbally attacked and, in one case, even physically assaulted. Haag believes that the campaign helped to reframe the act of feeding as a harmful, inhumane one. He admits that some people still feed and will continue to do so, but most of them are elderly, and few new recruits are adopting the habit. The decline, he believes, came from a social pressure to conform.

Haag's plan had another, frequently misunderstood component. With help from the city, he set up nine lofts in buildings around Basel that would house a small population of pigeons. The birds are not fed, simply given a place to stay. The researchers do remove some of the eggs the birds produce, and some people mistakenly assume that the lofts are part of the pigeon control strategy. But the lofts really serve as a research tool for Haag and also provide what he terms "an argument for the pigeon feeders." If the city willingly houses a few hundred pigeons but requires them to find their own food, it demonstrates that the city is not anti-pigeon and offers a better way to live with them. "You cannot only forbid," he said. "You must be an example, a positive example."

I visited one of the lofts with Ila Geigenfeind, a graduate student of Haag's with a cherubic face, long jet-black hair, and a strip of black eyeliner along her lashes. She led me by foot across the Rhine into a dense residential area, with cafés and shops along the streets and apartments above. "You can see this is pigeon area," Geigenfeind remarked, as a couple of birds fluttered overhead. Indeed, several pigeons were hanging out on ledges, windowsills, and spires of the old apartment buildings, and we found a pair building a nest above a shop awning. We arrived at an enormous stone church called St. Matthias. Geigenfeind pointed to a small open window in the high sloping roof of the church. A pigeon emerged from the blackness, took a look around, and flew off. She unlocked the heavy wooden door of the church and we climbed a winding sandstone staircase up into the bell tower of the church. We reached the metal workings of the church bell, and then climbed a narrow metal stair and crossed a walkway suspended above piles of wood pulp that had been thrown into the attic of the church for insulation. Finally we reached a room with a Plexiglas window, through which we could see about twenty pigeons gathered around a grid of lettered and numbered nest boxes. Giegenfeind and I donned white lab coats, green caps, and face masks before we entered the pigeons' lair—a necessary precaution, I soon learned, from the heavy and odorous air. She has the task of cleaning the loft once every couple of weeks, removing debris from the nests and cleaning dung from the floors with a long scraper.

At one end of the loft, the plywood floor sloped down to the small window that I had seen from below, at the entry of which birds gathered on their way in or out. Geigenfeind checked the nest boxes, which held terracotta dishes and several young pigeons at different ages. One fat squab sat on the floor with its parents, almost fully grown except for its large black beak. Two squabs, slightly younger, sat in a dish, a mess of half-opened black feathers. Geigenfeind took out one dish to show me a baby, ten days old, with a purple body covered in soft spikes that would soon bloom into feathers. Behind its purple beak we could see the slight indentation of its ears, not yet covered with down. Geigenfeind showed me the bulbous crop below the bird's head that held its delivery of food from its parents, and the small metal ring she had affixed to its foot the day before. She placed an even smaller baby, just a few days old, into my palm, where it nestled its head against my thumb. Its body was pink, perhaps three inches across, with half-closed eyes and a beak still tipped with the hard white egg-tooth it used to crack through its shell.

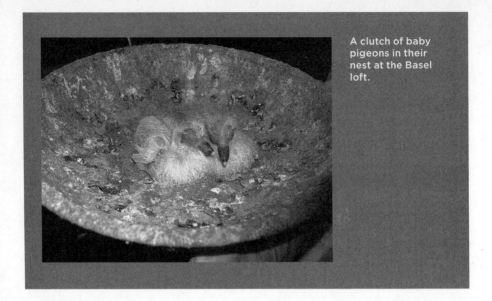

Geigenfeind does replace some of the eggs with plastic ones, but allows other youngsters to grow up here so that they will imprint on the loft and stay here as adults. Here the pigeons have protection from predators, and a nice high lookout to begin their journeys to find food during the day.

The lofts provide the rare opportunity to do careful research on feral pigeon ecology, like the growth of parasites that Geigenfeind studies, or even just answering questions about how the pigeons live. A former student of Haag's, Eva Rose, used GPS to follow the pigeons at some of the lofts, one of the only cases in which tracking has been used to uncover the movements of feral pigeons rather than homing pigeons. Even so, the study was conducted less out of curiosity about the birds than as a political tool—animal activists had argued that pigeons needed food because they could not fly well enough to go out of the city to find their own food. The flying abilities of pigeons are well known to people familiar with them, but less so to anyone who sees them wandering along a sidewalk. The study, Rose told me, was intended to show how much movement the birds are capable of.

The study found that pigeons are highly individual in the way they feed; each has a different strategy and a different set of favored feeding spots. They range as little as three-tenths of a kilometer to more than five kilometers, though most pigeons never fly beyond a kilometer or two from their homes. The most common strategy is to forage in streets and plazas near the loft, and pigeons usually fly outside the city only after they have checked out nearby sites. About half the pigeons also travel to the harbor, where they tend to gather on the roofs of the wooden warehouses and dive down to pick up spills from shipments of grain. Geigenfeind told me that pigeons often keep watch at bridges and roofs along the river for barges approaching the harbor. Females tend to fly longer distances, and are more likely to venture into agricultural areas outside the city or other large, predictable sources of food, perhaps because they cannot compete as well at sites where food is scarcer or need more energy for laying eggs.

Some pigeons adopt regular habits, visiting the same sites day after day. Others might hang out in street and squares one day and make a trip to the harbor another. Each strategy has trade-offs. The city streets are an unpredictable food source, but they are safer. The harbor and the fields are more predictable, but pigeons run the risk of being hunted by falcons and other predators on those journeys. The ideal food source would be both predictable and close, within the protection of the city. And that's exactly what the typical pigeon mother provides. A single person can shape the comings and goings of entire communities of pigeons.

Haag's documented success in Basel and his extensive knowledge of pigeons garnered him a bit of a celebrity in the pigeon-control world, and many people were interested in replicating his program. On a shelf in his office were several thick black binders where he collects correspondence, news clippings, and materials from projects in other places that purport to replicate Basel's approach. Because his work in Basel has been published in scientific journals, he has no rights to the idea. "They can take this idea, they can write what they want," he complained as he flipped through the alphabetically arranged binders.

Aachen, Germany. "The Basel model, it's called. But then they did something completely different."

Arad, Israel. Baden, Switzerland. Bath, England. Berlin.

"A photo of mine," he said, pointing to a poster. "They take all without any copyright."

He paused to open up another binder. We were into the "Ds" now. "Dusseldorf, it's a German city." It wasn't the use of his idea that he minded so much as the way his plan was invariably misconstrued along the way. "They feed the pigeons in the loft and say it is the Basel model."

It also irks him when other cities claim successes without actually tracking the pigeon population, something that takes a great deal of time and effort. It's impossible to know how many pigeons live in a given city—for every pigeon someone counts, Haag estimates, there may be five that are unseen. But what you can do is count the birds consistently—every week in the same place with the same person counting—and see whether the numbers rise or fall.

Haag doesn't necessarily believe that Basel's program is a model that should be carried all over the world. Each city needs to conduct its own scientific investigation to get at the underlying problem before attempting to fix it. But investing in this rational but expensive approach is something that few places outside of Basel have attempted. More likely is a quick-fix solution that will temporarily please taxpayers

But one town had been able to replicate Basel's success. The day before, the nearby city of Luzern had announced the success of its own pigeon-control program, which had been conducted over the past few years in consultation with Haag. I decided to visit Luzern to see how the program worked. Luzern is smaller than Basel but it receives a large number of tourists. Though Basel's center has a historic feel, Luzern captures the iconic Swiss setting that tourists want to see—impressive old hotels stacked along its river, narrow cobblestone streets, modern shopping, views of snow-capped mountains rising against the sky.

At the train station I met Monika Keller, the biologist and educator in charge of Luzern's pigeon-control program. Keller, a friendly woman with a tanned face and strawberry-blonde hair pulled into an easy ponytail, works for the government's natural resource agency. A few years ago, the city asked her agency to do something about all the pigeons, and mentioned Basel's success as a starting point. Since then, her team had been collaborating with Haag.

Keller and I walked along the picturesque riverside away from the plaza, where she gave me the tour of Luzern's pigeons. They liked to linger around the waterfront, resting along the top of the old wooden bridges. We saw a man tossing hunks of bread to swans sitting near the banks of the river—a couple of pigeons approached, looking for a share of the bounty, and he quickly tossed them a small piece. Keller explained that Luzern has to contend with all the tourists who feed birds as well as townsfolk. In front of the theater hall, we came to a large sign standing near the riverbank with a drawing of a large, lumpy pigeon; next to the drawing was a bar graph showing the pigeon population falling each year since 2001. Keller's team had just announced the success of their program: In five years, the estimated pigeon population had dropped from 7,000 to about 2,000.

Keller had organized this media campaign, which included other signs visibly posted along the river. Like Basel, Luzern now has its own pigeon loft as well; Keller pointed across the river where we could see the birds flying in and out of a tiny upper window in the elaborate Emmental roof of Luzern's Rathaus, a showpiece that houses city hall. From this prime location the birds have an easy view of the river and the train station. We crossed the river to the narrow street behind the Rathaus, entered a heavy door, and ascended the stairs of the building to the building's attic, a portion of which was sectioned off into a loft. In one corner, a Plexiglas panel showed a plywood-lined room similar to the loft at Basel, except that the nest boxes places along the walls were not numbered. A few pigeons sauntered across the floors—only thirty lived here so far, and there were many nests yet to be occupied.

Keller focused on education as well as persuasive media campaigns. In the center of the large attic she had set up a small museum with poster boards printed with information on the history and biology of pigeons. She leads tours for local citizens and schoolchildren—she points out behavior of pigeons in the streets, takes visitors to the loft, even brings them next door to the Picasso museum, where they can see photographs of Picasso's pigeons on his balcony at Cannes and look at his sculpture of a pigeon, its plumage suggested by blobs of blue paint. Her team even developed signs and flyers with recipes

Luzern's pigeons occupy prime real estate in the attic of the historical Rathaus.

A sign in Luzern illustrates its recent success at controlling pigeons and admonishes: "Now more than ever: pigeon-feeding is bad!"

Grafik des Monats
Taubenpopulation Stadt Luzern

Jetzt erst recht:
Tauben-Füttern ist schlecht!

using stale bread—since citizens here are charged a fee for every bag of garbage they throw out, they will sometimes feed their stale bread to birds rather than dump it. One of their signs, she told me, translated to "Bread brings shit." Out of necessity, Keller has focused most of her efforts on Luzern's citizens rather than its tourists, simply because of the challenges of crossing language barriers.

As in the Basel program, the loft in Luzern has nothing to do with reducing the pigeon population; Keller sees it as an educational tool. Her approach is not to convince people that pigeons are filthy or harmful but to develop an appreciation for the birds and to make the argument that feeding them is not in their best interests.

Later, in Keller's office in a nearby neighborhood, I asked about pigeon mothers in Luzern. Like Basel, Luzern has pigeon mothers, about thirty in Keller's estimate. She even knows their addresses (a policeman is one of the members of her team) and once a social worker even offered to make contact with them and open up a dialogue. Though the social worker talked with some of them, Keller isn't sure it made much of a difference. She is also unsure how much pigeon mothers contribute to the problem compared with tourists and other casual feeders—she estimates the regular feeders contribute between 50 and 60 percent of the pigeons' food, far less than Haag's estimate in Basel. "This is something we've discussed a lot," she said. On a beautiful day, everyone is outside and pigeons will see an increase in food, she explained, drawing an upward spike on a piece of paper. But spikes like these won't impact the entire population. The pigeon mothers, however, feed daily, so they may ultimately contribute more to pigeons' ecology than tourists ever will.

If pigeon mothers are a key part of the pigeon problem, they also seem to be an intractable one. Despite the success of their campaigns, both Basel and Luzern still have a small population of regular feeders. In both cities, some attempts were made to engage directly with the pigeon mothers using a mediator or social worker, but in both cases the mothers were not very responsive.

Keller believes that most of the town's pigeon mothers fall into the typical profile of an elderly, lonely, or even senile person. She and Haag both view pigeon-mothering as a cultural problem, a symptom of a fractured society. Haag showed me photographs of the apartment of one elderly man who died—the Spartan rooms were filthy and covered with dung from the birds he took in. These people reach out to animals because they are isolated and lack human contact, but they become so obsessive that their relationship with the birds becomes a sickness.

Of course, this vision doesn't account for Pigeon People or activists like Johanna Clearfield in New York—when I mentioned their activities to Keller, she exclaimed, "They're organized?" Sure, these pigeon activists are a bit on the fringe, but they hold jobs, maintain websites, and develop social relationships around their affinity for pigeons. They can hardly be dismissed as isolated.

I wondered whether Basel's approach could ever have the same kind of success in New York, or whether it was unique to a certain cultural environment. If human feeding shaped the ecology of pigeons, so did that environment's unique human culture. There is something Skinnerian in the way Haag's project engineers the social environment for the collective good. Social engineering of this sort is something that plays better in Switzerland than it does in the United States. I could easily spot signs that the cultural environment was very different from Boston. In Basel, the trains were always on time, and the bus stops were equipped with electronic signs to tell you exactly how long you'd have to wait for the next bus. I saw pedestrians waiting for walk signals when they easily could have made it across, and cars always stopped when someone approached a crosswalk. The streets, though winding, were well marked, and every tourist got a free pass on Basel's public transport system. The Rhine's waters, Geigenfeind told me as we crossed the bridge, were carefully monitored for cleanliness; we could see swimmers sunbathing on jetties along the banks. The Marketplatz is filled with fruit and vegetable vendors until about noon; by one-thirty it has been cleaned.

In short, Basel seemed to run in a fashion as orderly and smooth as the gears of a Swiss watch. It is the kind of place where people tend to obey the authority of a government that seeks rational solutions in citizens' best interests. It's not that Basel's citizens are blindly accepting of the pigeon-control program—at the recent yearly Carnival, one group chose to poke fun at it by dressing up as pigeon mothers, and carried an illustration of an old pigeon lady being watched by a policeman. But it does seem that the social pressure to conform was an effective tool here.

Which is why Haag's project, which seems to have succeeded so well in Switzerland, may be difficult to translate to another location. Haag himself gives the impression of someone who prefers an orderly, rational, and enlightened approach to problem-solving. Within his own laboratory and in Basel, he has been able to accomplish much. He rides to the university on an eerily silent electric scooter, teaches medical school classes, conducts his research with modest but adequate funding from the university and government, and takes coffee every day at ten and four with his graduate students so they can keep each other updated on their projects. He consults occasionally for cities interested in adopting his plans, and has published many articles and a book about pigeons.

He has had the good fortune to have support from a government that values scientific investigation and solutions. Walter Zeller, a representative at the veterinary authority in Basel, told me that the city appreciates Haag's work for giving it a sound scientific basis for its arguments. When I spoke with Zeller, I was simply impressed that someone in Basel's government actually thought about pigeons, as I had become accustomed to the institutional blind spot into which pigeons seemed to fall in the United States. In Boston, I had gone through a useless series of phone calls to find out if anyone was responsible for the city's pigeons. I was told by someone at the Massachusetts Department of Fish and Wildlife that they don't deal with pigeons since pigeons aren't really wildlife. I was referred to the environmental police number, where someone assured me that pigeons were a wild animal, therefore there was nothing to be done with them. The Humane Society might step in if pigeons were being improperly treated, but otherwise no one must enforce anything since no laws govern them—as the MassWildlife operator remarked, "there's an open season on them all year long." The Audubon Society is pigeon-neutral; the birds certainly don't fall within the institution's mandate to protect native birds, nor do they interfere with it, since there is no evidence that pigeons compete with native birds for territory.

Perhaps because of the science behind the project, or the unity between academics and government, or simply the Swiss mind-set, Basel was able to avoid the kinds of activist protest that mar many other pigeon-control programs. The situation is quite different even in neighboring Germany—Haag and Geigenfeind often expressed frustration at the strangely vocal animal-rights groups there who oppose their ideas. The two once attended a conference on pigeon control in Germany, but their talk was shouted down by protestors. They had to take their lunch in the kitchen.

If human attitudes toward pigeons shape the birds' ecology, they also must guide any sort of policy designed to control them. Not every population is going to be open to logic. Someone like Clearfield says that we must feed them because pigeons are here. Haag argues that pigeons are here *because* we feed them. Without human assistance, pigeons would still remain, but at more manageable densities. In national parks, it's easier to make the case that feeding animals is harmful—visitors are urged to respect the precious wildness of wild animals by resisting the urge to feed. But how do you make the case for a bird like the pigeon, which teeters on the edge of domestication?

Casual feeding—tossing a piece of bread to pigeons once in a while—has no effect on the birds' living situation. But regular feeding creates an entirely different situation, and an ethical problem. The birds can begin to structure their lives—where they nest, where they forage, and how often they reproduce—around their feeding schedule. They become dependent on their human feeders. These are the beginnings of domestication. And Haag argues that it's wrong for anyone to adopt one of the obligations of domestication without the other responsibilities that go with it, to feed the birds without giving them housing and cleaning up their mess.

When I stopped to talk to a man who fed the pigeons at Horticultural Hall, he told me that he loved animals and grew up on a farm in the West Indies where pigeons were kept along with other animals. When I suggested to him that feeding the pigeons could be a bad thing, he dismissed me with a wave of his hand. "They've got babies to feed!" he protested.

If pigeons retain some of the instincts of domestication, so do people. And ecological arguments are more difficult to understand than the bond between humans and animals. If domestication were a millennia-long relationship, it's no wonder the breakup is a little rocky. Now, when I thought about Raymond feeding pigeons at dawn, I saw that both he and the pigeons were enacting a ritual that had its roots deep in history. Perhaps the domestication of the pigeon began this way: with pigeons roosting in buildings and pigeon mothers beginning to care for them, long before the relationship became formalized.

Domestication may simply be the most natural way we interact with the other species in our environment—the species we like, at least. When we dislike something—cockroaches, rats, and in some cases pigeons—our instinct is to destroy it, to run it out of town, or at least our living rooms and backyards. But when an animal pleases us, our first instinct is to form a relationship. Children love to feed and play with wild things until they are told not to. Neutrality—letting wild things be wild—may actually be the most difficult thing for us to do.

I never had an urge to feed pigeons until I was watching the birds at the Pigeon Condo and noticed one that looked unusual. It was chasing two other pigeons around the small parking lot in the alley, and I soon realized it was a fledgling begging for food. The two adults refused, and continued to peck for food on the asphalt. The squab gave a few half-hearted pecks, but otherwise just stood near the two birds, waiting to be fed. It walked gingerly, lifting its feet slowly with each step, and occasionally fluffed its feathers out and hunched over a little. Eventually, the two adult birds flew off, and the squab slunk under the shadow of a concrete barrier below an exhaust pipe on one wall of the building.

When I returned some time later, the pigeon was still there. It still had a large black beak and scruffy black feathers that identified it as a youngster. When I approached, it flew up to a ledge near the nests, which reassured me that it could fly.

But the squab's condition never improved, and I realized it had been abandoned. I saw it chase adults a few times, but for the most part it just sat in the alley. It didn't seem to know how to be a pigeon;

sometimes it pecked the ground a little but never with the quick head-snap of a normal pigeon. Once I found it standing on top of a parked car—something I had never seen a pigeon do. It developed a limp that became more and more pronounced. When a car pulled into the alley and the squab was forced to fly, it executed a sloppy circle in the air and tried to land on a ledge on the second floor of the building, but only tumbled to the ground below.

Seeing a helpless creature moved me. My fiancé James and I returned to the alley one evening, bearing food for our pathetic little friend, aware that feeding it was as much to ease ourselves as for the pigeon. It had moved into a side alley, and sat still, its feathers erect and its body shaking. We tried to give it breadcrumbs and water in a spoon—in the darkness it didn't even try to move away and sucked haltingly at the food.

James, who once spent hours rescuing a mouse from a glue trap with rubbing alcohol and an Exacto knife, was ready to take the squab home and nurse it back to health. But taking a pigeon to our small apartment with a hungry cat didn't seem possible to me. And it was clear that in the remote case that this bird survived in our care, it would never return to the wild. In addition to a bad foot with abnormally long toes, it seemed to be blind in one eye—perhaps the defects that led its parents to give it up. It didn't seem right to try to prolong its painful life just to make ourselves feel better; only later did we decide that the best thing to do was to find someone to euthanize the poor thing. We left that night, and the image of that hunched, dark body, lumpy beak, glittering eye, and reaching toes, its helplessness and aloneness, all haunted me. Shortly after, James found it dead under a parked car.

When you start to look at pigeons, it's inevitable that you begin to see the sick and the injured ones. The birds whose toes got caught in string and became gangrenous and fell off. The birds with broken wings trying to flap coarsely to avoid a car. The squabs left by their parents. You might conclude that the city is a harsh place to be, but of course injury and disease happen everywhere in the wild, invisible to us. Most of the pigeon lovers I talked to first began their involvement with the birds when they encountered a vulnerable bird like this and felt compelled to help.

The next time we spotted an abandoned, injured squab in our neighborhood, we scooped it into a box and took it to an animal hospital where it was surely euthanized. Perhaps the vets silently cursed us for burdening them with a mere pigeon, but to me it was the best compromise between the disinterested ideal of wildlife and my own humanity.

And I knew that it was not just practical reasons that kept me from taking that first bird in, nor any notions about letting nature take its course. It was also the feeling that taking care of one sad little pigeon was a step down a slippery slope, and sympathetic as I was, I wasn't ready to become a pigeon mother.

12

Origin

Pigeons are not always the guests of humans. In 1850, the naturalist William Thompson, in his *Natural History of Ireland*, remarked at length about the rock doves living on coastal cliffs of the country, and his own experiences hunting them. These memories led him to reflect on the differences in scenery associated with rock doves—or rock pigeons as they are now known—and the related species ring dove (also called the wood pigeon). "The ring-dove is most at home in the lordly domain, rich in noble and majestic trees, the accumulated growth of centuries," he wrote. This was a species that nested in trees; it lived in the shelter of stately beeches along rolling hills, all scenes that Thompson associated with "the tender and the beautiful." Rock pigeons, in contrast, were always found mingling with "the stern and the sublime in nature." They lived beside great waterfalls, along cliffs that tumbled into the Atlantic and the Mediterranean, in caves and inlets of Scotland and the Greek Isles. The rock pigeon "has its abode in the gloomy caverns both of land and sea. How various are the scenes—nay, countries and climates—

brought vividly, with all their accompaniments, before the mind, by the sight of this handsome species!"

Thompson's remark is astounding when you consider how difficult it would be today to equate pigeons with the sublime. And yet the rugged scenery that Thompson extols is their origin. Hunting wild rock pigeons in their own domain was an adventure in Thompson's day; it involved hiking or boating to the base of sea cliffs or climbing along the windswept tops of them, sometimes scaling the cliffs to get a close enough shot. In hunting accounts, rock pigeons are elusive creatures.

Wild rock pigeons still exist, of course. Having traced the journey that pigeons took through domestication and back to the urban wild, I wanted to see some of the pigeons that still lived as they always had, untouched by human meddling. I convinced James, now my husband, to spend a portion of our honeymoon seeking out wild pigeons. After all, as Thompson noted, one can expect romantic views wherever the birds are found. We decided to visit Capo Caccia, Sardinia, the region where Richard Johnston had taken the specimen whose slight bones, clustered in their cardboard box, I had once weighed in my palm.

Capo Caccia lies to the north of the port town of Alghero, a peninsula of limestone that juts against the sea. The area was named for its popularity as a royal hunting ground and for the sport of hunting wild pigeons. Now the small area that is still an old forest is a park managed by the local forestry office. I was fortunate enough to make contact with Marco Apollonio, a biology professor at the University of Sassari in Sardinia, who persuaded two of his graduate students, Antonio Cossu and Laura Iacolina, to take us to a site where pigeons lived.

Cossu drove us in his car to the park, along a narrow dirt road that passed through a landscape of dry soil and pine trees, low bushes, and cacti. We parked in an empty dirt lot and the four of us traveled by foot along a path that led out of the woods and onto a rocky promontory above the ocean. The land sloped upward before it ended abruptly, falling to the sea in curtains of white rock that folded into grottos and canyons carved by waves. Across from where we stood on the shore was a gleaming island of chalky rock.

Even though the pigeons of Capo Caccia are wild and live in this seemingly remote setting, they still depend to some extent on people. There are berries and seeds along these cliffs that they can eat, but they prefer to travel to nearby agricultural fields where food is more concentrated and accessible. Cossu pointed out the saddle point between two hills across which the pigeons would fly to the fields, stopping on the way at a small man-made lake for water. Our hope, however, was that some of the pigeons had not left for the day and would be visible near their nests.

We picked our way along the cliff edge, climbing over rough rocks and through dense scrub bushes and junipers, stopping frequently to scan the cliff faces through binoculars. About fifty feet below us, the deep blue water of the Mediterranean roiled around the faces of the rock, kicking up pale foam. Swarms of seagulls circled the island and traveled over our cliffs. Everywhere we looked, we saw seagulls spinning, calling, pouring from the sky around fish catches, crossing overhead. But no pigeons.

Iacolina pointed to the empty sky, where we saw the crescent wings of a peregrine falcon soaring and then diving over an inlet near us, so we scuttled across the terrain in that direction, figuring that we might find the peregrine's prey. When we reached it, two pigeons flashed out of a small cave above a landfall of red earth at the top of a narrow canyon, but they quickly rounded a bend and disappeared.

I don't know what exactly I expected, but nothing in my experience had prepared me for this: To have to work to see a pigeon. I realized I had imagined the wild birds through the lens of my own associations with pigeons. Unconsciously I had assumed I simply had to show up and there they would be, sunning themselves on the cliffs just as they did on the sides of buildings. But in fact, these birds had no reason to loiter in such a stark landscape with peregrines roaming.

We sat down on the rocks and took turns scanning the cliff faces. Even oddly flattened through the lenses of my binoculars, the cliff walls seemed to hold beautiful nesting sites, inky pockets and fissures, overhangs where the sea disappeared into the rock. We began to grow hopeful at anything that looked like dung—but unfortunately, many of the rocks seemed to leach a white mineral that tantalizingly dripped out of cracks and holes.

We made our way back to where we had first begun traversing the cliff and hiked in the other direction a little further to the top of an uplifted bluff. To get there, we traveled through the juniper forest, pushing through brush, and emerged on higher ground with even more dramatic views, but still no pigeons. We sat on the rocks and ate the sandwiches we'd bought at a grocery store earlier. Someone noticed a pair flapping across the sky from the direction of the fields; they angled down to the cliff, turned a corner, and disappeared. We had been out for hours now and, disappointed, we headed back to the car. On the drive back, once in a while we spotted a pigeon or two flying overhead to and from the fields. Rolling hills gave way to eucalyptus forests and long flat beaches. When the red roofs of Alghero came into view, we saw plenty of pigeons everywhere.

Later, James and I wandered around the old part of the city, where feral pigeons lounged on windowsills and church turrets and wandered across cobblestone streets. On their perches they were often joined by a relative newcomer, the collared dove, another successful invader that rapidly colonized Europe in the first half of the twentieth century. But the rock pigeons may have lived here for centuries, living parallel lives to the ones at Capo Caccia. We even found several pigeons nesting in pockets in the rough brick walls that buffered the old city against the sea, a kind of imitation of their ancestral lifestyle.

In the 1980s, Richard Johnston had coauthored a paper in *The American Midland Naturalist* arguing that the wild pigeons of Europe were in danger of "genotypic extinction" because of encroaching feral pigeons. This sort of extinction may seem esoteric. After all, it's not as if pigeons are expected to disappear from the cliffs of Capo Caccia or any of the other places where they live. Instead, the loss is one of a genetic lineage; the birds themselves aren't threatened, but their unique gene pool is. A genome is a kind of historical record of a population's expe-

Antonio Cossu surveys the cliffs at Capo Caccia in vain for signs of wild pigeons.

rience. Wild pigeons have had a very different experience in the last few millennia than domestic or feral pigeons have. In truth, the two populations have probably been interbreeding for a long time in many locations. As early as the 1800s, a Scottish naturalist commented that it was difficult to get a pure specimen of wild pigeon, as domestic pigeons from nearby farms often mated with them. But they are separate enough that the populations are genetically still very different. Though considered one species, their genomes have had the opportunity to diverge considerably.

Geography is one of the main engines of diversity; when individuals live far apart or are kept separate by features in the landscape, they gradually become different from each other. When one variation of a species becomes invasive or is introduced into new territories where it can mix with others, the distinctness gets blended out. Connection has a way of flattening out differences.

Indeed, feral pigeons can be seen as part of a larger process threatening the diversity of life on the planet. They are a kind of avian franchise, a species that you can see in any major city. Just as a chain store like Wal-Mart can begin to replicate itself and invade new territories, pushing out other businesses, certain species have the ability to dominate the planet. Some scientists believe this process, as it occurs many times with many different species, is changing the organization of life on the planet just as globalization has shaped the world economy.

This concept is called "biotic homogenization," and it's still a nascent field of research. Scientists have studied the process of species invasions—how certain plants and animals are able to establish themselves in new places—for a long time. But the study of biotic homogenization looks at the result of the process, the effect of frequent species invasions on the uniqueness of different habitats.

Julian Olden, a researcher at University of Washington who has helped to promote the concept of biotic homogenization, said that regions that were historically very different in flora and fauna are now becoming more and more similar. And the source of this sameness, he said, is a group of "cosmopolitan" plants and animals that seem to show up everywhere. The vast majority of them have spread through some relationship with humans, whether they were directly used by people in agriculture or sport, or served as ornament, or inadvertently traveled on ships and airplanes. "They've piggybacked their way to different parts of the world," Olden explained. "As a result, regions that were once distinct are now very similar." In business, the spread of franchises and chains threatens not only the viability of small shops but the entire cultural uniqueness of different regions; in the same sense, biotic homogenization threatens to standardize the living world.

Chain stores would never threaten global domination if they weren't attractive to customers, and that's why they generate such controversy. Why shouldn't a more successful business survive? The same could be argued about species. Because of their enhanced ability to breed, Johnston said that if there were an outright race to reproduce between ferals and wild pigeons, "the feral birds would come to predominate throughout the world."

But in reality, wild and feral pigeons are not in any kind of a race. Though they may have some feeding sites in common, their habitats are so different that they are not in direct competition. But interbreeding can have untold consequences for the wild birds. "That primordial wild gene pool is being infiltrated by genetic elements that are characteristic of the feral birds," Johnston said, including genes honed from years of domestication. Whether those new elements can coexist with a wild lifestyle is unclear.

James and I decided to return to the cliffs alone the next day. The park's visitor hours were nine to four, which meant we couldn't try to catch the birds at dawn or dusk when they were more likely to congregate. But we arrived earlier in the morning, hoping that more of the birds would be near their nests. As we drove past fields just outside the forested land, I saw a large flock of pigeons feeding among the stubble.

Once back in the park, we walked out again to the edge of the cliffs, found a promontory with a good view, and just sat for a while. Many minutes passed, but then I saw two small flickers of gray below me. A pair of pigeons executed a swooping arc along the shoreline of the inlet below me, then headed off along the cliff face out of view. Though smaller and more compact, they looked like familiar pigeons; gray bodies and wings slashed with black, necks capped in iridescent feathers: the familiar blue bar coloring that many street pigeons have. The biggest difference was how fast they moved, flitting rather than pumping their wings, diving here and there, vanishing from view.

We crossed into the next canyon, where the birds had disappeared, and we glimpsed one pigeon flying out of the canyon, landing for a second or two on a rock above it, then flying off. The top of the canyon sloped down slightly to a steep rock crevice. I walked to the far side of the crevice and eased my way down the slope to an outcropping where I had a better view of the lower part of the cliff. I could see that it dropped into a wide cavern—at the bottom, the waves washed into a dark cave. The nest, then, might lie deep into the rock, completely hidden from view.

The white cliffs gleamed; I sat for a long time, hearing the surf against rock and trying to keep my eyes downward on the cave even though it brought a persistent feeling of vertigo. Once, I heard a familiar warble and a flap of wings, and a gray body shot past me from below, the closest I had seen one of the pigeons until now. We were having better luck. But still, when the birds did come, it was always just a flash of wings. They rarely stopped. We tried to take pictures, but only captured blurred chevrons against the rock or the ocean. They appeared mostly as pairs; later we saw a group of four fly in tight formation, darting around our inlet before heading down the coastline. Eventually they left the shore and flew toward the hills and fields, white underwings flashing against the sky.

By now, morning had become afternoon, and we finally decided to head back to the car. Though I had only caught glimpses of the birds, those glimpses were enlightening simply in how unexpected they were. After many hours spent under the daze of sun and white rock and the roar of waves, I could finally understand the connection that William Thompson had drawn between pigeons and the gorgeously stark scenery that evoked the sublime.

Later, I asked Emilio Baldaccini, a scientist at the University of Pisa who has been studying the Capo Caccia pigeons off and on for several decades, whether our experience was typical. He told me that indeed the birds were usually hidden within the rock walls, and he believed the caves in which they lived were networked to one another like the holes in a sponge. Once, his research team had tried to catch pigeons by throwing nets over the opening of a nest like the one we saw, only to find that the birds could fly out through other passages. He believes there may be three thousand birds living in the area, but it seems like there are fewer because they remain so hidden.

There were similarities between the wild and feral birds—their appearance, of course, but also their lifestyles. As Baldaccini's research has shown, the wild pigeons are commuters that fly several miles from their homes every day to find food, much like the pigeons of Basel, though the wild birds may cover more territory to reach their feeding spots. It's the same behavior that is cultivated, in more extreme form, in homing pigeons.

It was clear that each incarnation of the pigeon was linked to the other. But there was nothing in the elusive, bat-like birds I had just seen to indicate they were also capable of building nests on the surfaces of air conditioners, strolling nonchalantly through traffic, and loafing at busy street corners while begging for food. For the first time, I was able to perceive the great distance these birds had traveled—not just geographical distance but evolutionary distance as well.

Because much of this evolutionary change was caused by humans, the classic response is to see the genotypic extinction of wild pigeons as a tragedy. But when I asked Richard Johnston about the loss of the wild genotype, he said, "you might say it's being improved." After all, why should the wild birds be mourned if ferals are so much better at being pigeons? In their book, he and Janiga write, "if ferals ultimately prove to be the winners, humans could say, 'what a piece of work is the feral pigeon.' It has not, however, escaped our attention that the rest of the world looks at feral pigeons with considerably more ambivalence than we do."

Despite our natural ambivalence, there may be some use in learning about and even accepting the species that choose to live with us. In a recent paper in the journal *Conservation Biology*, a group of ecologists led by Robert Dunn at North Carolina State University make a case for what they call the "Pigeon Paradox," the idea that urban animals, even invasive ones, may paradoxically have an important role in conserving the Earth's species. "The urban jungle, with its many non-native species," they write, "may well be the breeding ground for future environmental action."

Dunn explained to me that for the first time in history, most people in the world live in urban areas, a fact that changes their relationship to nature. "To the extent that [if] they're going to make a connection to nature [it] is with the species in cities," he said. Dunn cites environmental psychologist Louise Chawla, who has argued that feeling connected to nature, particularly during childhood, makes people more likely to care about conservation and environmental issues. For those who don't have the ability or inclination to take camping trips and excursions into wilderness, the only chance they have to interact with the natural world is through the plants and animals in cities, many of which are cosmopolitan species brought by people. In their paper, Dunn and his colleagues argue that conservationists should focus more attention on restoring urban ecosystems and weigh "the costs and benefits linked to how we portray non-native and so-called pestiferous urban species."

Some invasive species are clearly harmful and deservedly shunned and eradicated. But urban animals in particular don't always compete with and harm native species; in many cases, they are taking over habitats that have already been made unliveable for other species by humans. Pigeons are able to create a natural habitat out of areas that seem hostile to animal life. In a way, they help to bring a bit of nature back into cities.

Much of the prejudice against urban animals comes not from environmentalists but from homeowners, landlords, and city officials who would rather not deal with them or have aesthetic objections. But ecologists and conservationists haven't been much more welcoming. "In a broader social context," Dunn said, "anything that's abundant is thought of as not very good." In the bird world, pigeons may have fallen farthest in human favor, but Canada geese and crows are also being seen as pests as they continue to swell in numbers. Dunn said that conservation biologists like him indirectly encourage these hierarchies; by definition, they only focus on the rarest species that need saving. Although it makes sense to funnel resources to the species that need help, Dunn worries about this "strange conservationist worldview we're imposing on the animal world." It creates the impression that species that are rare and threatened are better and more interesting, while devaluing the commonplace.

From a practical standpoint, Dunn also worries that fixating on the rarest species might ultimately backfire. He points out that people in major metropolitan areas like New York, Chicago, Boston, and Washington, DC, contribute the majority of dollars to conservation in the country. Ignoring the species that reside in large cities risks losing the attention of an important source of support for environmental causes.

Dunn believes that the tendency for science to ignore urban ecology is diminishing as more biologists themselves live in cities and begin to look for answers to questions around them. Like many people in his field, he grew up in a rural setting and came to his field out of a love of nature. His interest in urban wildlife only came later once he got a job in a city.

In their paper, Dunn and his colleagues write that "the lives of doves can open the door into a broader interest in wild nature." Some educational programs have already recognized the Pigeon Paradox and embraced urban wildlife. Project PigeonWatch, one of several urban bird programs at Cornell's Lab of Ornithology, enlists the public, and schoolchildren in particular, in collecting data about pigeons. The project, though it moves slowly, has a legitimate scientific goal: to try to understand why the colors and patterns of feral pigeons are so diverse and how color influences how pigeons choose their mates. Enticing children into science through an appreciation for the commonplace rather than the exotic has the potential to reverse some of the common hierarchies of science.

Some avid bird-watchers, the ones who keep checklists of every species they have seen in their lifetime, refuse to check off the rock pigeon until they see a "real" one in the wild. Now that I had seen wild pigeons for myself, I didn't feel that the ones back home in Boston were any less real. My experience in Sardinia was certainly much closer to what I thought of as "traditional" nature—the kind of experience I had seen on nature documentaries showing beautiful landscapes and the unusual animals that lived in them. Like Dunn, I grew up with easy access to the natural world, and I felt stifled when I later moved to Boston and spent my days in an environment I saw as ugly, sterile, and uninspiring. But in truth, I learned far more about pigeons' lives from standing in a dirty alley a few blocks from my apartment.

Documentaries are always filled with animals that are highly specialized—the ones with the most outrageous courtship rituals or the most improbable anatomies, living in the most extreme environments. Learning about these creatures and their exquisite adaptations helps people see the capabilities of evolution while catering to everyone's natural interest in seeing the biggest, the smallest, the rarest, the oddest. The pigeon is not the smartest bird, nor the fastest, nor the prettiest, and it is certainly not the rarest. But it is capable of so much. More specialized birds might illustrate the limits of evolution, but pigeons show us its breadth.

Seeing the wild pigeons at Capo Caccia helped me to understand the narrative of these birds that unfolded from remote origins to their current ubiquity. Whether the story is a tragedy, a comedy of error, or a tale of success depends on your perspective. But no matter how you choose to judge it, the story of pigeons' rise can offer something that studying a more elusive animal can't: a reflection of ourselves. Pigeons did not become "unnatural" through their association with us. Instead, our incredible influence on pigeons shows how much we truly are a force of nature. We create and destroy habitat, we shape genomes, we aid the worldwide movement of other species. And yet we seem disappointed and horrified when those plants and animals respond by adapting to our changes and thriving in them. All of life strives to survive. By thinking in false categories and separations, we miss seeing the survivors—even as they hover in multitudes above our street corners in the light of day.

Bibliography

1: The Pigeon's Progress

Jean Hansell's 1998 books *The Pigeon in History: Or, the Dove's Tale*, and *Doves and Dovecotes*, coauthored in 1988 with Peter Hansell, list the early symbolic history of pigeons and doves, and were a source of historical information for later chapters as well. Shakespeare's different meanings of "pigeon" and "dove" are explained in James Edmund Harting's *The Birds of Shakespeare*.

The official change of *Columba livia*'s common name appeared in 2003, and can be found in "Forty-Fourth Supplement to the American Ornithologists' Union Check-list of North American Birds" in *The Auk*. Derek Goodwin's observations about pigeons as a group come from *Pigeons and Doves of the World*, 2nd ed., published in 1983.

Richard Johnston's book *Feral Pigeons*, coauthored with Marian Janiga, will be mentioned several times, as it provided the inspiration for this book.

2: Invited Guests

For general background on invasive species, I relied on Jeffrey A. McNee-ley's introduction to *The Great Reshuffling: Human Dimensions of Invasive Alien Species*, which McNeeley edited. A more bird-specific account, including the failure of common quail to succeed in the United States, is in "The Ecology of Bird Introductions," a 2003 paper by Richard Duncan, Tim Blackburn, and Daniel Sol, published in the *Annual Review of Evolutionary and Ecological Systematics*. Helpful brief histories of the introduction of sparrows are on the Cornell Lab of Ornithology's website, All About Birds, http://www.birds.cornell.edu/AllAboutBirds.

In preparation for this chapter, I read Stephen Budiansky's 1992 book *The Covenant of the Wild: Why Animals Chose Domestication*, a nicely argued thesis about why domestication is a process of coevolution. These ideas resonated for me as I began thinking about the relationship between pigeons and people.

I have seen different estimates of when pigeons were first domesticated, so I used the date referred to in Johnston and Janiga's *Feral Pigeons*, as well as its account of the evidence of large-scale domestication in Egypt. For the history of pigeons in ancient cultures, the domestication of the pigeon, and an account of how people worked to keep pigeons in their dovecotes, I looked to Jean Hansell's *The Pigeon in History*. I also took the English translation of Comte de Buffon's quote about pigeons from this book. Many early general encyclopedias and natural history texts from the 1800s refer to pigeon-keeping; I found the entry in an 1839 edition of *The London Encyclopedia* helpful in explaining how pigeons were kept in dovecotes. An entry in the 1842 *Chamber's Information for the People* mentioned the difficulty of keeping birds in their dovecotes. The role of the dovecote or loft in Italian architecture is mentioned in Claudia Lazzaro's 1985 paper "Rustic Country House to Refined Farmhouse: The Evolution and Migration of an Architectural Form" in *The Journal of the Society of Architectural Historians*.

I found a few details about the use of pigeons as food in Roman times in Maguelonne Toussaint-Samat's *A History of Food*, a witty book that mentions how the Romans, in pursuit of poultry excellence, developed some cruel methods of fattening their birds: "Battery farming was a method they also used for pigeons, first breaking their legs."

I first read about pigeons coming to North America in Kim Todd's book *Tinkering with Eden: A Natural History of Exotics in America*. I later found many of the details she cited in a short 1952 account in *The Auk* by A. W. Schorger called "Introduction of the Domestic Pigeon." The history of the passenger pigeon has been recounted in many places, but I referred to the *Encyclopedia Smithsonian*'s detailed online entry.

Coming across Joan Thirsk's book *Alternative Agriculture: A History from the Black Death to the Present Day* was a nice breakthrough, because it detailed the rise and fall in popularity of dovecotes and tied it to a much larger economic process. For a sense of how pigeons were kept and bred in the early twentieth century, I read pigeon-breeding manuals of the day, such as Arthur Hazard's book from 1922, *Profitable Pigeon Breeding*. The handwritten note that I refer to came from a copy of this book in the Harvard University Depository. Also, Sheppard Knapp Haynes's *Practical Pigeon Production* from 1944 gave a history of squab raising, commercial squab farming in the United States, and discussed the profitability and advantages of keeping pigeons. Wendell Levi's *The Pigeon* contains a nice summary of the history of pigeon breeding and put the earlier pigeon manuals into an historical perspective as part of a larger "boom" in pigeon production in the United States. Levi's book includes a wealth of detail and advice about breeding birds, including photographs from his own pigeon farm showing how the birds were kept, killed, and packed for market.

A 2006 article by Dan Murphy on today's specialty game market on Meatingplace.com, a meat industry website, titled "Game is on for growing segment of specialty producers" quoted Tony Barwick. This, along with my later conversation with Barwick, helped me bring Thirsk's historical study of dovecotes to present-day pigeon farming.

One of the questions motivating this chapter was, if both chickens and pigeons were domesticated thousands of years ago, why is only one ubiquitous in supermarkets? I soon found that the question of whether pigeons "pay" has been an ongoing one. *The Commons Complaint*, published in 1611, was the earliest argument I found that pigeons were more costly than they were worth. The 1830 text I cite is *Baxter's Library of Agricultural and Horiticultural Knowledge* by John Baxter. Fancy pigeon breeders later also addressed the issue, including George Walton in *Essay on the Points and Properties of Fancy Pigeons* from 1876 quoted here.

For general information on pigeons and chickens as poultry, I referred to *The Science of Animal Husbandry*, 4th ed., edited by James Blakely and David

H. Blade. I also had a helpful email discussion with James Paul Thaxton, a professor of poultry science at Mississippi University, who explained some of the biological and reproductive factors that make raising pigeons and chickens so different. He recounted raising pigeons for experimental work, and came to a conclusion about pigeons similar to Buffon's: "In my view they are smarter than most household pets. They do not forget or forgive human interferences," he wrote. "When man tries to impose his 'systems of mass production' on the pigeon, these birds simply rebel."

3: Darwin's Metaphor

James A. Secord's insightful 1981 paper "Nature's Fancy: Charles Darwin and the Breeding of Pigeons" in *Isis* helped me decide to make Darwin's story a major part of the book, rather than just a side note. I was also fortunate that the paper gathered together Darwin's story with the historical background of pigeon fancying at the time, so I drew on information from the paper throughout this chapter. I also read Janet Browne's excellent biography of Darwin, and the end of the first volume, *Charles Darwin: Voyaging*, had a brief but very helpful account of Darwin's stint as pigeon-breeder.

For Darwin's own writings on pigeons, the beginning of *On the Origin of Species* was of course essential to this chapter, but his lengthy chapter on domestic pigeons in *The Variation of Animals and Plants Under Domestication* has a more detailed discussion of the different kinds of pigeons and their single origin. I also read through Darwin's collected letters during the time he was active with pigeons, beginning in spring 1855 and ending with two letters to William Tegetmeier in April of 1858, when he says, "In *about* six weeks' time I shall go over all my Pigeon M.S. & shall then dispose of all my Birds." He offered his remaining pigeons to Tegetmeier. "But I very much fear that few of my Birds, if any, will be worth your acceptance.— All my many crossed Birds I will kill, for I presume these cannot be worth anything to any body." *Darwin, the Norton Critical Edition* provided helpful biographical information and criticism of Darwin's works at the time, including the quote from Adam Sedgwick in this chapter.

It was particularly fun to search out some of the old "treatises" on pigeons that Darwin often refers to. John Moore's *A Treatise on Domestic Pigeons* was a well-known text at the time, having first been published in the early

1700s and reworked in later editions by other authors; I was able to look at an edition from 1765. I also found a text on pigeons by William Tegetmeier called *Pigeons: Their Structure, Varieties, Habits, and Management*. Tegetmeier assisted Darwin greatly in learning about pigeons, though this particular text was from 1868, after Darwin's work with them.

4: Hopeful Monsters

The rich historical texts on fancy pigeons again were a wonderful resource for this chapter. In addition to the treatises by John Moore and William Tegetmeier cited above, I read E. S. Star's 1886 book *The Breeding of Fancy Pigeons*. By this time, Darwin himself had undoubtedly influenced the way fanciers saw their birds, and Star begins the book with this quote from Darwin: "I know it as an art and a mystery." Star even discusses some of Darwin's experiments with pigeons, but he was not convinced that all fancy pigeons come from the common "blue rock" pigeon. I also read George Walton's slim 1876 volume, *Essay on the Points and Properties of Fancy Pigeons*.

John Moore's original *Treatise* had contained a section devoted specifically to the Almond Tumbler. In 1851, John M. Eaton published *A Treatise on the Art of Breeding and Managing the Almond Tumbler*, which drew from Moore's original work and included a detailed description of what the perfect Almond Tumbler should look like. For more modern information on the breeds, I looked to Levi's *The Pigeon*, which, despite being published decades ago (my copy was from 1977), is still considered the Bible on pigeons today. For a more theoretical take on the fancier's enterprise, see Walker van Riper's 1932 paper "Aesthetic Notions in Animal Breeding," published in *The Quarterly Review of Biology*.

I visited pigeon shows in Connecticut, New York, Rhode Island, and Massachusetts before finally going to the National Pigeon Show. I'm glad I did so, as the first experience walking into a pigeon show is quite disorienting, and it took time to reach the point at which I could even begin to look at the birds in the same way their owners did. It also helped that when I reached the Grand Nationals I already knew people competing, particularly a friendly and helpful group from Pennsylvania. Though the other shows are not mentioned in this chapter, my conversations with breeders there were incredibly helpful in preparing me for this chapter.

A conversation with biologist Trevor Price linked fancy pigeon breeding to the process of sexual selection: a paper by him along these lines titled "Domesticated Birds as a Model for the Genetics of Speciation by Sexual Selection" was published in 2002 in the journal *Genetica*. John Fondon's work with pigeons is, as of yet, still unpublished. Jill Helms outlines her case for using fancy pigeons as laboratory subjects in a paper called "The Origin of Species–Specific Facial Morphology: The Proof Is in the Pigeon," published in 2007 in *Integrative and Comparative Biology*. Inspired by a comment from Fondon, I also bought a copy of Stephen Green-Armytage's book of photography, *Extraordinary Pigeons*. I highly recommend it for seeing fancy pigeons in their full beauty and freakish splendor.

5: Homing

I learned about pigeons' flying abilities from a conversation with Bret Tobalske. Jean Hansell nicely summarizes the entire history of pigeons as messengers, including their symbolic use, in *The Pigeon in History: Or, the Dove's Tale*, as well as in *Doves and Dovecotes*, coauthored with Peter Hansell. As in many other chapters, Wendell Levi's *The Pigeon* was an important source of information; in this case, for the history of homing pigeons in the military, including a detailed account of known pigeon "soldiers" and medal recipients. I also found a useful account of pigeons in the British stock exchange in "The Stock Exchange," *Old and New London: Volume 1*, published in 1878, which can be accessed through British History Online (http://www.british-history.ac.uk/).

There is an enormous scientific literature on pigeon homing, and it remains an active field. Two reviews I found particularly useful were "Pigeon Homing: Observations, Experiments and Confusions" by Charles Walcott, published in the *Journal of Experimental Biology* in 1996; "Avian Navigation: From Historical to Modern Concepts" by Roswitha and Wolfgang Wiltschko, published in *Animal Behavior* in 2003. I also spoke with Dora Biro at Oxford University about her work using GPS to track pigeons' homing; her paper "From Compromise to Leadership in Pigeon Homing" was published in *Current Biology* in 2006.

Before I planned to write this book, I visited the South Shore Flyers, a local pigeon racing club, as part of a multimedia project. That club, incidentally, was later featured in a *New Yorker* article by Susan Orlean, on February 13, 2006. My conversations with the club members helped me get a sense of the lengths to which racers will go to win, and the strategies they use to train and race the birds.

6: Hunt and Peck

Robert D. Nye's 1992 book *The Legacy of B. F. Skinner: Concepts and Perspectives, Controversies and Misunderstandings*, gives a clear overview of Skinner's ideas and attempts to disentangle what he said from what has been said about him. For biographical details and chronology, I used Daniel W. Bjork's 1993 biography, *B. F. Skinner: A Life*. Skinner, however, was a prolific writer so I was able to find plenty about his work with pigeons in his writing. The second volume of his autobiography, *The Shaping of a Behaviorist* (1979), has an account of his time working at Project Pigeon and later at Harvard. A more specific and lively account can be found in his paper "Pigeons in a Pelican," published in 1960 in *American Psychologist*. I also benefited from an outside narrative of this time in James H. Capshew's paper "Engineering Behavior: Project Pigeon, World War II, and the Conditioning of B. F. Skinner," published in 1993 in *Technology and Culture*. During a visit to the B. F. Skinner archives at Harvard University, I was also able to read the original typewritten reports on Project Pigeon, one prepared for General Mills in 1944, and the other for the Office of Naval Research in 1952. The set of Polaroids in this chapter comes from the later report.

The *Journal of the Experimental Analysis of Behavior* published a series of papers in 2002 about the Harvard Pigeon Lab, some of which contained anecdotes and impressions from former students and provide a nice sense of the excitement surrounding behaviorist research in those days. Skinner gives an account of his discovery of the power of reinforcement with pigeons in a paper entitled "Reinforcement Today" published in 1958 in *American Psychologist*. His demonstration of "superstition" in the pigeon was first published in the *Journal of Experimental Psychology* in 1948. The later examples of using pigeons as "quality-control" monitors came from *Control of Human Behavior*, a 1966 book edited by John Ulrich et al. Writer Beth Azar cited the Coast Guard pigeon project in a 2002 article in the American Psychological Association's *AMA Monitor* called "Pigeons as Baggage Screeners, Rats as Rescuers."

It seemed impossible to write about Skinner without confronting the rift that occurred between behaviorists and cognitive scientists. I spoke with William Baum, who was a student in the Harvard Pigeon Lab and is firmly in the behaviorist camp, about why he feels behaviorism is still relevant. And in addition to my visit with Robert Cook, I spoke at length with Donald Blough, a psychologist at Yale who also trained at Harvard and continues to use pigeons in his research, who found behaviorism unsatisfying and took a cognitive approach. Brett Gibson, a cognitive psychologist at University of New Hampshire also helped orient me to current cognitive research with pigeons. For a detailed argument for why strict behaviorism should go the way of the dinosaurs, read Stephen S. Ilardi and David Feldman's 2001 manifesto "The Cognitive Neuroscience Paradigm: A Unifying Metatheoretical Framework for the Science and Practice of Clinical Psychology," in the *Journal of Clinical Psychology*. Ultimately, I realized that the theoretical framework didn't matter all that much from a pigeon's perspective, since cognitive approaches today often draw from the same methods developed by Skinner and his colleagues. Perhaps, though, the tasks the pigeons perform are more interesting.

7: Escape of the Superdoves

This chapter relied a great deal on a section about what is known of the origin of feral pigeons in Johnston and Janiga's book, which lists many of the specific cases of street pigeon sightings mentioned here. That book alludes to the "freeing" of pigeons during the French Revolution, and I found a few more details in a paper by John Markoff titled "Violence, Emancipation, and Democracy: The Countryside and the French Revolution," published in 1995 in *The American Historical Review*. Charles Townsend's account of pigeons in Boston was published in 1915 in *The Auk*; it helped to illuminate the history of pigeons where I lived.

In researching this book, I wondered about when the "street pigeon" became a recognizable city inhabitant. Thanks to online databases like Google Books, I was able to search for random mentions of pigeons in a wide swath of literature. All my searching did help me to get a sense of how people saw pigeons, but I never truly found a satisfying answer. In the *New York Times* historical database I found an article from 1882 titled "Park Commissioners' Ways" that mentioned pigeons in Central Park. An 1872 article called "The

Chattering Sparrow" argued that sparrows were encroaching on the territory of pigeons. "Only a few years ago pigeons fed in the streets, among the horse-droppings, without danger of attack." The author concluded that "pigeon life in the City must be a good deal of a burden." Margaret Welch's *The Book of Nature* provided background on early natural history in the United States, which helped me understand how the focus on native species came from an early sense of nationalism. The example of pigeons as pests in Turku, Finland, came from "Urban Biodiversity in Local Newspapers: A Historical Perspective," published in *Biodiversity and Conservation* in 2001, and was very helpful in showing me that today's attitude towards pigeons has a history at least a century long.

The work on sparrows by Richard Johnston and Robert Selander was published in *Science* in 1964. Most of Johnston's pigeon work is summarized in *Feral Pigeons*. A few other papers were useful: "Variation in Size and Shape in Pigeons, *Columba livia*," published in the *Wilson Bulletin* in 1990, and "Geographic Variation of Size in Feral Pigeons," published in *The Auk* in 1994.

Louis Lefebvre has an interesting collection of papers on social behavior in pigeons, but for this chapter I focused on his work on species invasions and innovation. "Behavioral Flexibility and Invasion Success in Birds" was published in *Animal Behavior* in 2002; "Brains, Innovations and Evolution in Birds and Primates" was published in 2004 in *Brain, Behavior and Evolution*. A 2007 paper, "Social Learning and Innovation Are Positively Correlated in Pigeons (*Columba livia*)" in *Animal Cognition* is his most recent work in progress. Though Lefebvre here is presented as a bit of a foil, my several fascinating conversations with him about pigeons helped inspire me from the beginning.

8: A Squab Is Born

This chapter draws heavily on the detailed description of feral pigeons' courtship rituals, mating, reproductive habits, nest-building, and parenting behavior in *Feral Pigeons*. Many of the studies cited came from other researchers throughout the world, some of which I looked at, but the book does a nice job of summarizing what is known about how pigeons live. Johnston also published a paper titled "Nonrandom Mating in Feral Pigeons" with Steven G. Johnson in a 1989 issue of *The Auk*. I should note that, looking through a database of ornithological literature for references to pigeons, I found brief descriptions of different sexual behavior, including an episode of homosexuality and an attempt by a male pigeon to mate with a mallard. And Johnston notes that pigeons will sometimes mate with a bird other than their partner. So, as always, one should never assume that the heterosexual monogamous ideal is always followed. The wire pigeon nest mentioned here was from a brief 1997 account with photo, "An Unusual Rock Dove Nest" in *The Auk* by Robert L. Paterson Jr. For information on peregrines I relied on conversations with Tom French and Bill Davis, and educational materials from the Massachusetts Division of Fisheries and Wildlife.

9: The Urban Habitat

The "pigeon census" of Milan was published in a 2002 paper titled "Effects of Building Features on Density and Flock Distribution of Feral Pigeon *Columba livia* var. domestica in an Urban Environment" by Roberto Sacchi et al., in the *Canadian Journal of Zoology*. The influence of socioeconomics on bird diversity was first noted by Stephanie J. Melles in a paper in *Urban Habitats* in 2005 called "Urban Bird Diversity as an Indicator of Human Social Diversity and Economic Inequality in Vancouver, British Columbia." For background on the urban ecology of pigeons and birds in general, I spoke with Charles Nilon at the University of Missouri, Robert Blair at the University of Minnesota, John Hadidian of the Humane Society of the United States, and Karen Purcell of Project PigeonWatch at Cornell University. Much more has been written on the negative effects of urbanization than on its benefits. Early in my research I also read a nice 2001 scientific book, *Avian Ecology and Conservation in an Urbanizing World*, edited by John M. Marzluff, Reed Bowman, and Roarke Donnelly, which helped orient me to some of these issues as they relate to birds.

10: Defining Pigeons

Most of the events in this chapter happened much earlier than the others, though it appears later in my narrative; Al Streit and others continue to be active in New York City, most recently trying to block a proposed thousand dollar fine for feeding pigeons. To find out about the pigeon activists of New

York, I monitored the conversations in the Pijn People Yahoo! group listserv as well as attended one of their meetings. In addition to conversations with Al Streit, Johanna Clearfield, and Anthony Pilney, I spoke at length with Carol Cavers, another pigeon activist, and Bob of Bird Operations Busted (BOB), a somewhat secretive man who is conducting surveillance on pigeon netters in the city. The account of pigeons on Wall Street is from a *New York Times* article in 1935, "Haughty Pigeons Ousted in Wall St." This was the first really serious case of conflict that I found in the *New York Times* database. From there it devolved; the later article from 1959 entitled "Friends and Foes of *Columba Livia*," authored by David Dempsey, describes a conflict over pigeons that could have been written today.

11: Pigeon Mothers

The plan to eradicate rock pigeons in the Galápagos was outlined by R. Brand Phillips et al. in a 2003 paper in *Noticias Galápagos*, "Feral Rock Doves in the Galápagos Islands: Biological and Economic Threats." I spoke with Bryan Milstead and Jabier Zabala at the Charles Darwin Foundation about how pigeons were completely exterminated from the island; apparently, pigeon activists have not migrated to the Galápagos, because the project seems to have been completed without fuss.

Daniel Haag-Wackernagel systematically exposed pigeons to various deterrent systems, and found the birds could surmount all of them if they really wanted to, and that those that cause pain were no more effective. See "Behavioral Responses of the Feral Pigeon (*Columbidea*) to Deterring Systems," in *Folia Zoologica*, 2000. A study in Italy that found a "pigeon pill" not to be a very effective control method was published in *Wildlife Research* in 2007 by D. Giunchi et al.

Haag-Wackernagel has published papers on the ecology of Basel's pigeons. The first paper that caught my attention was a fairly recent one using GPS to track feral pigeons—the first tracking study I had found that didn't use homing pigeons or wild rock pigeons. I spoke with him and first author Eva Rose about the paper, "Practical Use of GPS-localization of Feral Pigeons *Columba livia* in the Urban Environment," published in *Ibis* in 2006. His pigeon-control success in Basel achieved a worldwide audience with a brief letter in *Nature* in 1993 entitled "Street Pigeons in Basel." A more detailed account can be found in "Regulation of the Street Pigeon in Basel" in a 1995 issue of *Wildlife Society Bulletin*. He was also kind enough to provide me with a copy of his book, *Die Taube*, and though I can't read German, it's fascinating for the wealth of images on pigeons in history.

I corresponded by email with Nancy Severance at the Audubon Society about their official position on pigeons. Severance said that Audubon is "not generally involved in pigeon issues because they are a non-native species." But she added that "as we have no evidence that pigeons compete with native birds or have any negative effects on native birds, we do not promote pigeon control."

12: Origin

William Thompson's description of rock doves and their habitats appeared in *The Natural History of Ireland*, vol. II, published in 1850. I found many other brief mentions of wild rock doves in natural history books and encyclopedias of the time. William Yarell, a friend of Darwin's, talks about the known geographical reach of the birds in *A History of British Birds*, published in 1843. Richard Johnston's paper "European Populations of the Rock Dove *Columba livia* and Genotypic Extinction" appeared in *American Midland Naturalist* in 1988. Emilio Baldaccini and colleagues have published a few papers on wild rock doves. "Foraging Flights of Wild Rock Doves (*Columba l. livia*): A Spatio-Temporal Analysis," published in the *Italian Journal of Zoology* in 2000, gave me some information about the Capo Caccia pigeons and their habits, which Baldaccini expanded upon in our conversations by phone and email. Marco Apollonio also gave me information by phone about Capo Caccia's history and the habits of its resident pigeons.

Julian Olden and colleagues nicely summarize the biotic homogenization argument in "Ecological and Evolutionary Consequences of Biotic Homogenization" in a 2004 opinion paper in *Trends in Ecology and Evolution*. "The Pigeon Paradox: Dependence of Global Conservation on Urban Nature," by Robert Dunn et al., was published in *Conservation Biology* in 2006. Anyone interested in learning more about pigeons or participating in citizen science research should visit Project PigeonWatch's website at http://www.birds.cornell.edu/pigeonwatch.